NEWTON AT THE BAT

NEWTON AT THE BAT

THE SCIENCE IN SPORTS

edited by

Eric W. Schrier and William F. Allman

CHARLES SCRIBNER'S SONS

New York

Copyright © *1984 The American Association*
for the Advancement of Science

Library of Congress Cataloging in Publication Data

Main entry under title:
Newton at the bat.
  Includes index.
  1. Sports—Physiological aspects.   2. Force and
energy.   I. Schrier, Eric.   II. Allman, William F.
RC1235.N48   1984   796'.01'5   84-1299
ISBN 0-684-18130-4

3 5 7 9 11 13 15 17 19   F/C   20 18 16 14 12 10 8 6 4

Printed in the United States of America.

All essays in this book appeared, in slightly different form, in Science 80, Science 81,
Science 82, Science 83, *or* Science 84 *magazine.*

Contents

Part III
THE BODY

Part IV
FORM

Preface

ERIC W. SCHRIER

Grant Giske, the bullhorn-throated coach of the ramshackle Trident Swim Club of Palo Alto, California, always taught his kids two things. First, watch the starter awhile before your race. If you're smart, Coach Giske would say, you can pick up the rhythm of his call: SWIMMERS TAKE YOUR MARKS (count one, two, three) BANG. That way, when your turn came, you could get off to a rolling start.

Good advice, but it was nothing compared to the second thing. When you're coming into the wall at the finish, he would say, always splash the timers. They don't really want to get wet, and if the splash arrives before you do, they'll jump back early, clicking their stopwatches as they go.

I guess there wasn't a whole lot Coach Giske figured he could do about the middle of my race. And he was probably right. I was a scrawny, no-talent nine-year-old. But boy, could I start and finish.

By the time I got to college, someone in swimming officialdom had wised up. Where there once was the predictable bang of a starting gun, now electronic beeps at each block started the race. (The gun method had been unfair to swimmers in the lane farthest from the starter, given that they heard the sound from the gun five-hundredths of a second later than swimmers in the first lane.) And where there once were timid timers, now there was a rubber pad on the wall of each lane recording the finishing touches to the nearest hundredth of a second—and it couldn't care less about getting wet.

That's when I really started worrying about the middle of my race: What was the best path for my hand? What muscles should I be strengthening with weights? How much should I swim each day during practice? There were answers to these questions back then, but they were based almost exclusively on intuition. This was 1970, still a few years before sports physiologists and sports medicine journals and biomechanics laboratories were invented. Computer-enhanced images of tennis serves and pole vaults were nowhere in sight. The entire town of Eugene, Oregon, may have been running then, but the rest of the country still called the shoes they ran around in "sneakers" and didn't have the slightest idea how long a marathon was.

There are better answers to my questions now, and the world-record times in swimming show it. Where coaches used to have only intuition, now they have intuition with a little physiology or physics on the side. Not just in the pool but on the diamond, the track, the fairway, the court. There is a best arc for a jump shot. There are ways to make a bicycle go faster. There are certain foods you should eat the night before a marathon. There is a best position for speed when you're hurtling down a mountain at ninety miles per hour.

In his book, *The Sweet Spot in Time*, John Jerome talks about that moment of physical perfection—when the ball jumps off the bat, a bike sweeps into a high-speed turn, or the diver knifes into the water—that we all experience at one time or another. It is a feeling that occurs when our movements come together at just the right time in just the right place. It is the grace that we see often in a sport's greatest athletes and that we are sometimes lucky enough to feel ourselves, whether we're throwing a Frisbee or simply running down the street.

The science of sports is partly devoted to finding this metaphoric sweet spot, breaking it down, and devising ways an athlete can more easily find it again. The best path for my hand through the water, for example, turns out to be an S-curve. The best time for a marksman to pull the trigger is between heartbeats. Ski jumpers will fly farther if they stretch their body over their skis in the shape of an airplane wing.

Knowing what the body must do and doing it are, of course, two different things. Knowing that the knuckleball makes one-third of a rotation on its way to the plate won't help you hit it. And knowing how the muscle fibers in your shoulder slide past each other when you lift weights won't help you press more of it. In fact, after listening to an explanation of what a baseball does after it's thrown or how a tennis ball caroms off the racquet, most athletes will say something like, "I never think about it much, I just do it." Professional athletes are usually more interested in what science can tell them about how to avoid injuries or come back from them. Or what new equipment might give them that extra edge. It is because of such desires that we get carbohydrate loading and arthroscopic knee surgery, Prince tennis racquets at Wimbledon and venturi chassis at Indy.

For most of us, though, the science in sports really serves our curiosity more than our knees. If anything, the answers to the little questions we ask when we're watching the World Series or skipping a rock across a pond have a lot to do with our appreciation of the game itself. And that is as it should be. This book, then, is for anyone who has ever wondered if a curve ball really drops just before it gets to the batter. Or stared at that little white ball on its tee and asked why it has dimples. Does it really make any difference which running shoe you buy? Will a couple of beers help your dart game? And why, for God's sake, does the boomerang keep coming back?

Acknowledgments

The editors would like to thank Margaret R. Brodnick,
Mary Elisabeth Challinor, Margo Crabtree,
and the *Science 84* staff for their help.

Part I

BALLS AND OTHER FLYING OBJECTS

Pitching Rainbows: The Untold Physics of the Curve Ball

WILLIAM F. ALLMAN

"A real good curve is slower than a fastball and breaks straight down. It doesn't begin to bite until about twenty feet from home plate."
—Eddie Murray, first baseman

"The seams create a vacuum. It's going to go so far before the pitch starts to grab and pull down. Vacuum, drag, gravity, velocity, centrifugal force, it's all of them. On a curve ball, the vacuum is the main thing."
—Jerry Koosman, pitcher

"It's moving more in the second half than in the first, that's what hitters call a break. But a physicist would say it's a perfect circle."
—Robert Watts, mechanical engineer

For more than a hundred years it has plagued major league hitters. It has been called many things: drop ball, hammer, out-shoot, downer, and hook. Its common name is the curve. Several times in its history scientists claimed it was merely an optical illusion. Baseball players, who knew better all along, told the scientists to stick to their books. Finally conceding that the ball does curve after all, scientists discovered the forces behind the mysterious motion, but not before the players had made a few misguided attempts at baseball aerodynamics. The scientists told them to stick to their bats.

3

Ray Miller puts topspin on a curve by twisting his fingers over the top of the ball. He also turns his hand out during his follow-through, a motion he did not know occurred until he saw the strobe pictures of his delivery. "When this is all over," said Miller, who is the pitching coach for the Baltimore Orioles, "I'll probably find out I don't know what the hell I'm talking about." (© Gordon Gahan/PRISM and Science 82*/stroboscopic lighting by C. E. Miller/MIT)*

You'd think a quiet peace would have finally settled onto this lab and locker room debate. It hasn't. Sure, everybody agrees that the curve ball curves, but nobody can agree on *where* it curves.

"A curve ball comes straight in," says Montreal Expos pinch hitter Terry Crowley, "then about four or five feet from home plate it breaks straight down." Sandy Koufax's curve, hitters say, looked as if it rolled off a table.

Physicists disagree, saying that the force on the ball is constant, so the ball must be curving constantly. According to Robert Watts, a mechanical engineer at Tulane University, if gravity did not affect the ball and nothing got in the way, a curve would travel in a complete circle. "The pitcher," he says, "could turn around and catch the pitch he just threw."

Most major league pitchers throw curve balls, even those who are known for their blistering fastballs. The curve is thrown a little slower, which upsets the batter's timing, and its motion can be very effective when the pitch is mixed in with a little smoke. "It doesn't look like the ball starts breaking until about ten feet away," says Orioles catcher Joe Nolan, "which can make a hitter flinch 'cause he's never too sure if the thing's a fast ball or a breaking ball. It's a matter of hanging in there. Some players can, some players can't. The ones that can't have other jobs by now."

It was a little easier for hitters to keep their jobs in baseball's early years. The batter, stepping up to the plate, could request that a pitch be low or high, and the pitcher was obliged to throw it that way—without raising his hand above his hip. This delivery method made early baseball a hitter's game. In 1869 the Cincinnati Red Stockings beat the Cincinnati Buckeyes by ninety-five runs.

By the turn of the century, pitchers, finally allowed to throw overhand, were making things considerably tougher. J. Franklin "Home Run" Baker led the league in home runs four seasons in a row, racking up a total of only thirty-nine homers in the process. In 1919 Babe Ruth set a major league record by hitting twenty-nine balls out of the park. But the next year the pitcher's practice of spitting on, scuffing, nicking, and otherwise tampering with the baseball, which made the ball harder to hit and sluggish when it was, became illegal. That year Ruth hit fifty-four.

When pitchers could no longer fold, spindle, and mutilate the ball, they had to rely more on spin. The curve ball had been striking out batters since the Civil War, despite claims by some that the pitch did not really curve at all. In an 1870 demonstration Fred Goldsmith reportedly threw a sidearm curve at three upright poles arranged in a straight line. According to Henry Chadwick, an early

This strobe-lit view over Scott McGregor's shoulder demonstrates that most major league curves have little sideways movement. "A true curve ball breaks over the top," says Orioles outfielder John Lowenstein. "It breaks straight down. That's what makes it so difficult to hit. The bat has to come through the vertical arc of the ball." (© C. E. Miller/MIT and Science 82)

historian of the game, the crowd "cheered lustily" as the ball traveled to the right of the first, the left of the second, and the right of
the third. That would seem to have laid the matter to rest, but *Life*
magazine, which commissioned strobe light photographs of curves
in 1941, disagreed. The "evidence fails to show the existence of a
curve," the authors wrote after examining the photos, and "raises
once more the possibility that this standby of baseball is after all
only an optical illusion." In keeping with their rivalry, *Look* magazine also took strobe photographs and concluded that the ball *did*
curve.

To throw a curve, a pitcher usually holds his index and middle
fingers close together along a seam of the ball, and as he brings the
ball past his ear he snaps his fingers over the top—somewhat like
turning a doorknob on a door that is edge-on—and lets it roll over
his middle fingers. This gives the ball topspin. Of course, a pitcher
can make the ball curve in any direction by altering the angle of
the spin. For example, a sidearm delivery gives the ball sideways
spin, making it curve horizontally.

Though curves that break sideways were popular with baseball's
founding fathers, most of today's major league pitchers try to
make the ball curve downward. Their reasoning is simple: A baseball bat is essentially an eight-inch-long, three-inch-wide cylinder
with a handle. Lateral motion in a curve just brings the ball to
another part of the bat. But if the pitch moves down, it can cause a
hitter to miscalculate the height of his swing, leading to a pop-up
or grounder if not a complete miss.

The curve ball curves because of the 216 stitches of red cotton
that hold the ball's cowhide cover together. According to the late
Lyman J. Briggs, former director of the National Bureau of Stan-
dards, the stitches on the ball pull a thin layer of air around with
them as they spin. Expanding on earlier work by the noted aerody-
namicist Igor Sikorsky, Briggs found that on a curve ball that has
topspin, this spinning air layer causes more air to flow around the
bottom than the top. The bottom air travels faster than its upper
counterpart. The more rapidly air flows over a surface, the less
pressure it exerts on that surface. Thus the air pressure is lower
on the bottom of the ball than on the top, and the ball is pushed
down. Briggs found that a major league curve ball, which rotates
some eighteen times in the half second it takes to make its way to
the plate, can curve as much as seventeen and a half inches.

Just where that curve takes place has been disputed for more
than a century. The 1877 edition of *Spaulding's Baseball Guide* says,
"It is not necessary to waste space to prove that the ball can be

These four pictures were made by flashing a bright strobe light 120 times a second as the ball traveled through a darkened wind tunnel. The two photographs on the top recorded the first half of the ball's flight, and the two on the bottom captured the second half. On each pair, Jim Walton, an engineer at General Motors' Biomedical Science Department, compared

images of the ball on one photograph to their sister images made simultaneously on the other. By noting the position of the image pairs relative to the Ping-Pong balls, Walton could derive the exact coordinates of the ball's flight. (© C. E. Miller/MIT and Science 82*)*

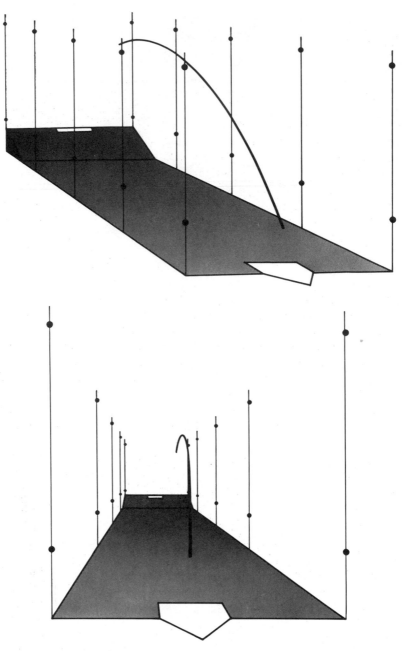

Once photo analyst Jim Walton gathered the data from the photographs and fed it into a graphics computer, he could reproduce McGregor's curve from any viewpoint in any direction. When viewed from the stands, top, the curve seems much tamer than when looking down the "tunnel," bottom, as a right-handed batter would. (adapted by Flash Fleischer)

made to change its course to right or left. The only real question is, whether the change in direction is instantaneous or gradual—that is, whether it is an angle or a curve."

"Do you want to know about a curve?" asks veteran catcher Carlton Fisk. "A curve ball goes *vwoosh!* (he sticks his hand straight out), then *vwoosh!* (the hand goes straight down)." Physicists argue that any force the pitcher puts on the ball begins to work as soon as it leaves his hand. "There's no way the force is going to suddenly take effect ten feet from the plate," says Watts.

"The break is not an optical illusion," says Orioles outfielder John Lowenstein. "All a physicist has to do is stand up there with a bat to convince himself." Watts may not have batted against a major league pitcher, but he has pitched semipro baseball in Baton Rouge, Louisiana, and he has published a technical paper on the aerodynamics of a knuckleball.

In the summer of 1982, Watts and two other scientists, Charles Miller and Jim Walton, took a look at the flight of the curve. Miller is an electrical engineer at the Massachusetts Institute of Technology's Strobe Alley and an expert at taking high-speed photographs. Walton, an engineer from General Motors Laboratories, admits that what he knows about baseball "you could write on the back of a postage stamp." He is from Britain, where they play a different game. But he does know a lot about analyzing high-speed photographs to determine the precise path of an object in motion.

To eliminate problems with perspective, Miller constructed a "tunnel" eight feet wide and eight feet high, stretching from the mound to the plate. The walls of the tunnel were marked by Ping-Pong balls hung from the ceiling. As Orioles pitcher Scott McGregor and pitching coach Ray Miller threw, one pair of cameras recorded the first half of the ball's flight and another pair, the second half.

In effect, each pair of cameras produced a stereo image that allowed Walton, whose computer program did the sophisticated trigonometry, to use the precise positions of the Ping-Pong balls to track the ball's path through the tunnel to within a tenth of an inch. Watts used the data to calculate the forces that make the curve ball curve.

Watts found that the curving force on the ball is constant, making the ball travel in a smooth arc from the pitcher's hand to the catcher. If it traveled sideways and there were no gravity, McGregor's curve ball would form a circle with a more than 2,000-foot diameter, circumnavigating Baltimore's Memorial Stadium and parking lot.

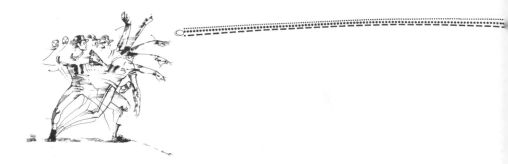

The anatomy of a major league curve: This schematic diagram shows the essential elements of a curve ball. When a pitch is released horizontally with topspin, the curving force pushes the ball down. The small dotted line represents the motion imparted by spin only and not due to gravity. On McGregor's curve, this force makes the ball move more than a foot. The large dotted line is the arc that would be formed if the ball fell by the force

The curving force does not alter the ball's speed but gravity does. A pitch that takes four-tenths of a second to reach the catcher, for example, drops only six inches due to gravity in the first half of flight, but it drops more than two feet in the second half. Still, the motion is gradual; there is no sharp break.

A curve looks like it's breaking because the hitter is standing near the circle formed by the curving ball. Think of a train traveling at a constant speed on a circular track. If a viewer stands at the center of the circle, the train appears to be moving at a constant rate. If the viewer stands near the perimeter of the circle, however, the train at first appears to be traveling in a straight line as it comes toward him and then seems to move sideways as it passes him and heads upward. This illusion is what causes the apparent break in a curve ball. A downward breaking curve ball is released more or less horizontally, its motion nearly in line with the batter's view of the pitcher, but as it approaches a hitter, its path becomes more vertical and the ball's motion can be more easily seen.

"If you take a picture of the curve and show it to a scientist, he'll say 'I told you so,'" says Watts, "and if you take it to a batter, he'll

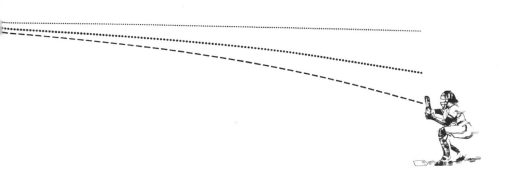

of gravity alone. (The ball slows down a little because of drag, but the effect is slight.) Because the curving force is in the same direction as gravity, the curve ball "falls" faster, shown by the dashed line. Thus, in a sense, hitters are right. Though there is no sharp "break" in its trajectory, both the gravitational and curving force on the ball make it move more in the last half of its flight than in the first half. (© Dean Williams)

say 'I told you so,' too. The curving force is producing an absolute circular path, it's not breaking. But if you ask the batter, he'll say he sees it curve only a few tenths of a foot in the first half of its flight and curving a foot and a half in the second. It's still a circular arc, but as far as he's concerned, it's falling off a table. It's all a matter of perspective. The physicist is looking at it from the side. The batter is looking down the tube. But both are actually saying the same thing."

Some baseball players knew better all along. Scott McGregor has always felt the curve was an arc, as does Oriole Rich Dauer: "A curve is a big swoop. It's like you had a rock on a string and you swung it. It goes out and it curves in. Like you drew it with a compass. It's always the same rainbow."

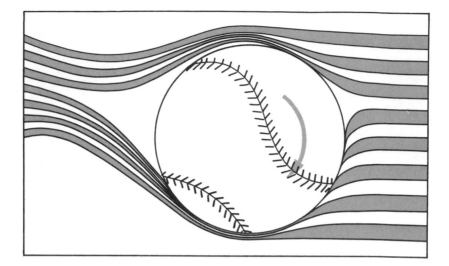

Spinning clockwise in a wind tunnel, a baseball disrupts the flow of smoke streams injected from the right. The stitches on the ball pull a thin layer of air around with them as they spin. This spinning layer causes more air to flow around the bottom than the top. The bottom air travels faster and thus makes the air pressure on the bottom of the ball lower, pushing the ball down. (Top– © Flash Fleischer. Bottom– © Prof. F. N. M. Brown of University of Notre Dame)

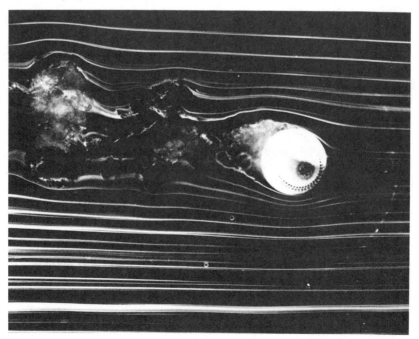

What Makes a
Knuckleball Dance?

WILLIAM F. ALLMAN

When a knuckleball leaves the pitcher's hand, it is not unlike the game of baseball itself: slow, static, and delivered with deft precision. But also like the game, the results of the delivery are quite the opposite: The ball's flight to the plate is unpredictable, quick to change, and dramatic when it does. And it is also a joy to watch—if you don't have a bat in your hand.

Yankee hitter Bobby Murcer once said that hitting a knuckleball is like "trying to eat Jell-O with chopsticks. Sometimes you might get a piece, but most of the time you get hungry."

When a knuckleballer is throwing well, the pitch is almost impossible to hit. Phil Niekro, the knuckleballing ace, was 17–4 for the Atlanta Braves in 1982 at the ripe old age of forty-three. His younger brother Joe, a knuckleballer with the Houston Astros, also won seventeen games. Sent to the minors in the early 1970s, Joe learned his brother's specialty and came back to become one of the best pitchers in baseball.

Most knuckleball pitchers hold the ball with the fingertips, not the knuckles, and throw it while keeping the wrist stiff. The pitch travels about fifty miles per hour, slower than curves and fastballs. But what it lacks in speed it makes up in movement. It darts and weaves, dips and sails, with the seeming capriciousness of George Steinbrenner contemplating his team's managerial post. The ball is so erratic that catchers working with knuckleballers use an over-

To throw a knuckleball, Phil Niekro holds the ball with his fingertips and throws with a locked wrist. (© Manny Rubio)

sized glove. Even then they have problems. "The best way to catch a knuckleball," according to former catcher Bob Uecker, "is to wait until the ball stops rolling and then pick it up."

Some struggling pitchers, lured to the pitch for salvation, instead turn to salivation. The knuckleball is hard to learn and control; it is much simpler—and more effective—to apply a little moisture to the ball. Though banned more than sixty years ago, the spitball and its modern counterpart, the Vaseline-ball, are still used by some pitchers. The ball is slickened and released like a pumpkin seed squirted between the fingers. The result is a pitch that moves like a knuckleball but at the speed of a fastball, and often leaves even the best hitters twisting in the wind.

The salient feature of the knuckleball and spitball is their spin— or lack thereof. Phil Niekro says he tries to throw the ball so it has no spin at all. Baseball wisdom holds that a ball without spin is subject to the whimsy of any passing breeze arising while the ball is traveling to the plate. The lack of spin also helps disguise the ball's motion. "It throws off the batter," says Niekro. "If he can see a spin, he knows which way the ball will break. If there's no spin, he doesn't know where it's going."

But as always with baseball, there are people who think they know better. Mechanical engineer Robert Watts, for example, says the best knuckleballs are those that spin a quarter revolution on the way to the plate. The key to the ball's movement, he says, is not in the passing of an occasional breeze but in the devilry of the air rushing over the ball as it flies.

The stitches on a baseball disrupt the flow of air around the ball, creating a force that makes the knuckleball change direction. In a wind tunnel at Tulane University, Watts measured this force as he turned different faces of a baseball into the wind. He found that some orientations of the ball produced a force that pushed it in one direction, but if the ball were rotated a few degrees, the force acted in the opposite direction. Thus a knuckleball that turns slowly on its way to the plate can actually change direction several times.

Other physicists think the knuckler's movement may be due to drag. If a pitcher throws the ball fast enough, the air near the front of the ball will become turbulent, split, and follow the contour around the ball. But if the pitch travels more slowly, a peculiar thing happens. At a particular velocity the turbulent air will switch to a smooth, or laminar, flow. This smoothly flowing air does not move around the ball completely but breaks away from the surface near the back, forming a wake like that of a ship. The wake creates

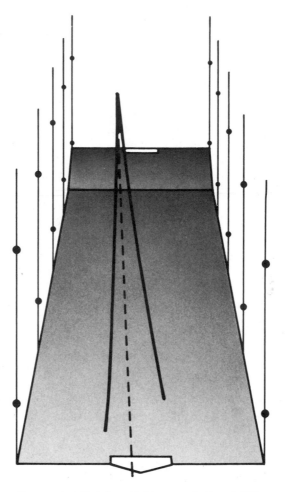

Two pitches, thrown by Orioles pitching coach Ray Miller and superimposed on each other by computer, show what makes a knuckleball so effective. Though thrown with the same motion and initial trajectory (as shown by the dotted line), one pitch moved left and the other moved right. (adapted by Flash Fleischer)

drag on the ball, sometimes as much as four times that of a faster pitch, and the ball slows down.

Some physicists think that when a knuckleballer talks of "finding his groove," he is adjusting his pitch so that it begins at a speed just above this aerodynamic switchover. As the ball moves toward the plate, it slows down and its drag goes up, making the ball suddenly drop.

The speed at which the drag crisis occurs on a sphere, however, depends on the roughness of the ball and its spin. And nobody seems to know just how the combination of stitch and spin affects the aerodynamics of a baseball.

Watts doesn't think that the drag crisis is involved in the knuckleball's buck and wing. "I ran the wind in my tunnel up to 80 miles per hour," he says. "I didn't get any signs of a big drop in drag. My guess is that the crisis occurs at about 250 miles an hour. And if you can throw a baseball 250 miles an hour, you don't need to worry about learning to throw a knuckleball." And while the switch from turbulent to smooth airflow may occur suddenly on some spheres, no one knows what effect the stitches of a baseball have on bringing it about. One theoretical calculation suggests that for a baseball the change in drag is gradual; the ball's velocity must drop by forty miles per hour before the drag goes up. Most pitches slow only about five to ten miles per hour on the way to the plate.

One piece of evidence to add to the controversy was gathered during *Science 82*'s curve ball experiment [see pages 3–13, "Pitching Rainbows: The Untold Physics of the Curve Ball"]. The knucklers and spitters thrown by Baltimore Orioles pitching coach Ray Miller curved both left and right, but none slowed dramatically. The next step, says Watts, is to put the ball in a larger wind tunnel and increase the air speed until the drag crisis occurs.

It seems that just about the only people satisfied with the vagaries of the knuckleball are the people who throw it for a living— though a few batters, too, seem unperturbed. Former slugger Richie Allen, for example, claimed the knuckler wasn't a problem for him. "I never worry about it," Allen once said. "I just take my three swings and go and sit on the bench. I'm afraid if I even think about hitting it, I'll mess up my swing for life."

Baseball's Dirty Tricks

STEPHEN S. HALL

"The tradition of professional baseball has been agreeably free of charity," Heywood Broun noted back in 1923. "The rule is, 'Do anything you can get away with.'"

Sixty years later, it may be added, charity still does not begin at home . . . or at first or on the pitcher's mound. Whether it is increasing friction at the grip of a moving object (i.e., using pine tar on the handle of a bat) or creating an aerodynamic irregularity on a moving sphere (i.e., throwing a scuffed-up or cut baseball), all players are amateur practitioners of applied physics when it comes to the tricks of the trade.

Batters, of course, believe they have the toughest job in the game. A variation of a millimeter or two at the point where bat meets ball can make the difference between a Rod Carew and a Rod Kanehl. A batter facing an eighty-five-mile-per-hour fastball on its fifty-six-foot journey from pitcher's hand to home plate has only about forty-five-hundredths of a second to decide if and when to swing.

That is also about how long the chalk lines of the batter's box remain intact at the beginning of the game. With the boundaries blurred, some batters stand further from the mound, gaining perhaps three inches of patience. Those three inches translate into an extra two-thousandths of a second to look over the pitch. Two milliseconds.

To a few, that advantage is too slight to matter and they step, figuratively, even further over the line. Players have been known to drill a half-inch diameter hole down from the fat end of the bat—anywhere from eight to fourteen inches—and cap it or fill it with cork. Some feel that if cork is stuffed into the core, the bat might transfer energy to the ball better—and also not sound too hollow if a suspecting umpire or opposing catcher decided to tap it on the ground. But the people at Hillerich and Bradsby, makers of the "Louisville Slugger" bat, say corking a bat doesn't do anything except make it more likely to break.

"They're wrong, I can guarantee you that," counters former Detroit Tiger first baseman Norm Cash. Thoroughly unrepentant, Cash admits using a corked bat en route to winning the 1961 American League batting title with a .361 average (along with 41 home runs and 132 runs batted in). Cash, like most ball players a self-confessed nonphysicist, explains the advantage of corked bats with a layman's simplicity and clarity: "The faster you can swing the bat, the more you can hit the ball."

Cash has the law on his side in this case—Newtonian laws of motion, to be exact. "It's the same principle as when an ice skater wants to go into a spin and pulls the body in to spin faster," explains Peter Brancazio, a professor of physics at Brooklyn College. "The greater the distance of the mass from the axis of rotation, the harder it is to rotate." Top-heavy bats are harder to turn around. And bat speed, both hitters and physicists agree, is more important than mass when it comes to making good contact with the ball.

The trend in major league baseball is toward lighter bats with thin handles and thick barrels—the type preferred by Rod Carew of the California Angels. It is also a nightmare for manufacturers, because thin-handled bats break easily. "The problem," says Rex Bradley of Hillerich and Bradsby, "is that the good Lord doesn't make timber that good." Yet the Carew-style bat is a step closer to the ideal baseball bat, which in fact would be a golf club—at least in an ideal world in which the ball was stationary. The head of a golf club, like the "sweet spot" of a bat (about four inches down from the end of the barrel), is where the best contact is made; the thin shaft is just there to move the head through the air. Likewise with Carew's thin handle.

There are also tales of bats with flattened sides (Nellie Fox reputedly used one), bats whose hollowed innards were filled with a free-floating dollop of mercury, bats fortified around the midsection with tacks, and even a bat stuffed like sausage with Super

Balls (such a bat reportedly broke open on American Leaguer Graig Nettles in 1974.) According to Paul Kirkpatrick, a retired Stanford physicist, "That's going in the wrong direction. The harder the material, the more energy will be returned to the ball after a collision. If you put a bat in a vise and bang on it, it won't vibrate long; it absorbs the force. But a steel bar will ring a few seconds because very little of the collision is absorbed. If a bat were made entirely of cork the only thing it would be good for is bunting."

A perfectly legal but vanishing tradition is the old clubhouse soupbone. Players used to expend considerable elbow grease rubbing the "sweet spot" of their bats on a soupbone to harden the surface (today's practitioners use soda pop bottles).

"Yeah, I used to bone 'em," says former slugger Frank Howard. "I'd spend two hours boning 'em and rubbing 'em and then the first time up, I'd hit one off the end or on the handle and break the damn bat."

"You could do things with bats years ago that you can't do anymore," adds former Yankee shortstop Tony Kubek, now an announcer for NBC Sports. "We used to take an ice pick and carve out the dark grain, which was the softer part of the wood. Then you'd put pine tar in the grooves, let it harden, and resand it." This supposedly gave the bat a harder surface and put more sock into a swing, although even Kubek concedes that the main advantage may have been psychological.

Rex Bradley recalls hearing of cases where the grooves in a bat were carved out but never filled in—"like scoring on a golf club," he says. The idea is that a grooved bat, like a grooved golf club, imparts backspin when it makes contact with the bottom of the ball. Others maintain that the direction of the bat at impact is much more important in generating backspin than grooves in the bat. But there seems to be no question that backspin helps a fly ball. "I imagine it could add 20 to 30 feet to a 350-foot hit," says Brancazio, and that could easily be the difference between a long out and a home run.

To take away that advantage, of course, the home team could always toss its supply of baseballs into the freezer. This is exactly what the good-pitch, no-hit Chicago White Sox used to do— according to Tony Kubek and others—when the Yankees and other hard-hitting teams came to town in the late 1950s. A frozen ball shows no spunk: The all-important coefficient of restitution— the ability of an object (such as a baseball) to rebound after a collision with another hard object (such as a bat)—goes down with temperature. This is because the core of each baseball is a combination of cork and rubber, and rubber becomes less bouncy when the temperature is lowered.

Talk of resilience and spunk invariably provokes discussion of the ball's liveliness. There is quite a lively discussion going on right now, in fact, between Rawlings, which manufactures all major league baseballs, and MacGregor Sporting Goods, which would like to. Like many corporate discussions these days, this one is taking place in court. In an antitrust suit against Rawlings, MacGregor charges among other things that the Rawlings ball fails to meet major league specifications. Major league baseballs, for example are supposed to weigh between five and five and one quarter ounces and have a coefficient of restitution of between .514 and .578 (this determined after a collision when the ball is traveling about fifty-eight miles per hour). For a softball traveling at fifty-five miles per hour, by contrast, the coefficient of restitution is .400, while for a Super Ball it is .850.

MacGregor hired Richard Brandt, a physicist at New York University, to test about fifty Rawlings baseballs. He found that 20 percent of the balls had coefficients of restitution higher than the .578 limit—in one case as high as .607. Brandt adds, "the balls are supposed to have a circumference between nine and nine and one quarter inches, but almost half of the balls were too small." Both these factors individually contribute to a livelier baseball, and in combination they can make "a significant difference," according to Brandt. "If you compare today's baseballs with ones from the past, the old ones were 25 to 30 percent less lively. What that means is that if you went back to 1927, when Babe Ruth hit sixty home runs, and you used the modern ball, Ruth would have hit eighty home runs."

Rawlings maintains that Brandt uses an unorthodox method of measuring the coefficient of restitution, hitting the balls with a flat metal plate instead of shooting them out of an air cannon. Brandt responds that "any freshman physics student could tell you that the results will be the same." Brandt, in fact, tested the baseballs both ways, and says the results were exactly the same.

Until they hear otherwise, Brandt's findings give pitchers one more reason to justify their time-tested mischief. In a sport where Hall of Famer Whitey Ford once admitted to throwing a doctored pitch in an Old Timer's game, it is obvious that choirboys and winners don't sit on the same bench. Baseball outlawed the spitball in 1920, but it is the sport's version of the fifty-five-mile-an-hour speed limit: It has been estimated that one-third to one half of all pitchers in the majors throw doctored balls.

And a pitcher doesn't have to go to his mouth to make the ball dance. Any irregularity in the cowhide surface—whether it is cut, scuffed, shined, or freighted with mud—changes the pattern of airflow around the baseball as it makes its journey to the plate. All that's needed is to make one side rougher than the other, and this can be accomplished by making one side smoother than usual.

"In the [1982] World Series we saw something that is technically legal: Bruce Sutter (the St. Louis relief pitcher) rubbing the ball on his pants," says Tony Kubek. "Well, that's what we used to call a 'shiner'—a dry spitball." Others, according to one American League veteran, will use the less legal sandpaper—often on the official label, where it is more difficult for umpires to pick up. A scuff ball is thrown somewhat like a whiffle ball: The pitcher puts the rough side away from the way he wants it to break. For example, to make the pitch curve right, the scuff is put on the left and thrown overhand.

There may be a tiny consolation in all this for hitters who face scuff balls and other ballistic indignities: If they make good contact, the ball may actually travel a bit further. The boundary layer of a scuffed ball experiences turbulence, and this in turn reduces the size of the wake behind the ball, thus reducing drag. In short, a roughed-up ball is aerodynamically more efficient than a perfect sphere. "Is that right?" asked a wide-eyed Detroit slugger Darrell Evans when Brancazio passed on the good news. "Usually we try to get the scuffed ones out of the game, but . . ."

In Search of
the Perfect Jump Shot

THOMAS H. MAUGH II

In his final year of professional play Rick Barry of the Houston
Rockets was the most successful free-throw shooter in the National Basketball Association, completing 93.5 percent of his attempts. His career average of 90.0 percent is also the highest in the
NBA. This might not be particularly unusual except that Barry
shot his free throws in a style different from that of virtually
everyone else. Disdaining the conventional one-handed shot,
Barry grasped the ball with both hands, swung it between his legs,
and lofted it in a high arc toward the hoop. Despite his skill at the
free-throw line, however, Barry was rarely among the top percentage shooters for field goals. The difference, suggests Brooklyn
College physicist Peter J. Brancazio, may have more to do with the
physics of Barry's unusual free-throw style than with his innate
abilities as a basketball player.

A self-professed basketball fanatic who has been playing regularly since he was a teenager, Brancazio began studying the physics of basketball a few years ago when he was unable to find
anything about it in the literature. He candidly admits that his
study was motivated by a desire to find some way to compensate
for his own stature (he is five feet ten inches tall in his sneakers),
his inability to jump more than eight inches off the floor, and his
advancing age. Brancazio's findings transformed some of what was

25

previously thought of as simply coach's intuition into a hardwood science.

Most of Brancazio's approach can be adequately explained only in the language of mathematics. Fortunately, his conclusions can be expressed in relatively simple English. Take the case of Barry, for example. Brancazio has shown that a ball spinning backward loses more energy when it hits the rim of the basket or the ground than does a ball spinning forward or one with no spin at all. The greater the amount of backspin, the more energy the ball loses. Therefore, when a ball with a high backspin hits the rim, it doesn't bounce off wildly but "dies" there, with a much greater chance of falling through the hoop.

Barry's underhand shot puts a great deal of backspin on it. Spectators have often noted, in fact, that when his shots hit the inside of the rim, they generally fall through. "I found a quote reading through some coaching books," Brancazio says, "where Auerbach [general manager of the Boston Celtics] says 'Backspin helps the shot to be lucky and fall in.' Well, it's not luck. There really is a physical reason why that happens."

Brancazio also calculated the trajectories for certain shots given their launch speed, height, and angle of release. Since a basketball is about half the size of the basket, a shot does not have to travel in a perfect arc for a score. A shooter thus has a margin for error in launching his shot, and Brancazio found that this margin for error increases when the ball is released at a greater height above the floor—"one more advantage, unfortunately, for tall people."

The Brooklyn professor also found that for any specific distance and launching height, there is a specific shooting angle, a "golden arch," so to speak, that not only allows the greatest margin of error but also requires the smallest launch speed to reach the basket. "This is an advantage," Brancazio says, "because the less force you put on it, the easier it is to get the shot off. If you're trying to take a jump shot over a tough defense, you want to get the ball away quickly without having to put a lot of force on it."

In general, from any spot on the court, a high, arching shot is much better than a flat one. And the closer the shooter is to the basket, the higher that arc should be. A bad launching angle can override other considerations, even backspin. Wilt Chamberlain, for example, was one of the most notoriously poor free-throw shooters of the modern era, despite his other prodigious talents. He too tried the Barry style of free-throw shooting but without any significant improvement in his average. "The problem was," says Brancazio, "that he was throwing line drives. He simply was

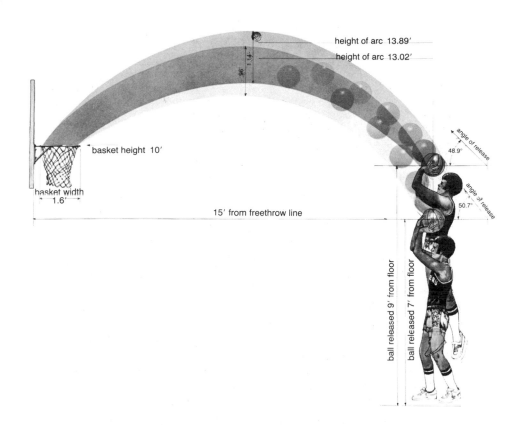

As the height at which a shot is released increases, so do the chances of scoring. A shot from the floor can travel any arc within the lower shaded area and score. But if the shooter jumps two feet, the width of this "scoring band," shown in the sweeping arc on top, increases by 18 percent. (© Tom Bookwalter/Hellman Design Associates)

not arching his shots enough."

To test his results, Brancazio filmed Brooklyn College students taking set and jump shots at various distances from the basket. Analyzing the film, he found that successful shots were being launched at angles very close to the minimum-speed angle for the given height of release and distance. It seems, he says, that "the better shooters naturally learn to shoot at the minimum-speed angle without realizing it." Analyzing his own game, he found that his launch angle was generally not great enough. Increasing it improved his shooting percentage.

Brancazio has not disseminated his findings widely, but he argues that they could be most useful to basketball coaches working with young players. "The most useful thing, I think, is teaching kids to shoot. I coach a kids' team of twelve- to fourteen-year-olds, and I found that my results were pretty useful, once I worked them out. They were trying to shoot the ball on a line at the basket, and I kept teaching them to try and shoot the ball high and make it come down into the basket. They were pretty impressed and, you know, it's hard to impress twelve-year-old kids."

A Slice of Golf

ANTHONY CHASE

There may never have been a time when people didn't work themselves into sweaty fits playing with balls. The first balls, possibly used as ritualized substitutes for live prey, were made of animal skins wrapped around stone or lumps of wood. Some were stuffed with human hair, bird feathers, dirt, or, as in the Aztec Indian culture, made with vulcanized rubber. Although the first ball games were simple forms of catch, early cave paintings show that the throwers soon turned into hitters, knocking balls around with sticks and clubs. And once there were hitters, the craftsmen soon appeared, fashioning better and better clubs to hit better balls.

Few sports have come farther from their lowly origins than golf. But the respectability and technological refinement now associated with the sport came slowly. In 1457, for example, the king of Scotland, concerned that a golf-crazed populace was wasting time thrashing balls around rather than practicing archery during a time of war with England, published a royal decree in which he demanded that "Golfe be utterly cryed downe!" But by the turn of the century, peace was made, and the armies of archers returned to their balls and clubs. Ironically, the very weapons craftsmen whose businesses first suffered because of the golf bug eventually benefited from the demand for clubs, mallets, and balls. No one was better suited than the bowyers and fletchers to master the intricacies of club shaft flexibility and the niceties of a snug leather

29

grip. Their influence transformed the golf club from a farm implement to a work of precision, making the game more accessible and thereby increasing its popularity.

Technological innovation was not limited to clubs. The earliest golf balls, made of carved boxwood, were replaced in the seventeenth century by a ball called the featherie. The featherie, manufactured by a technique of Roman origin, consisted of a small outer casing of untanned cowhide that was stitched to the desired shape with a hole left in the side, then turned inside out, leaving the stitches on the inside. The hole was stuffed with as many boiled feathers as would fill a top hat. Then the hole was sewn shut, and the feathers were left to dry. When the leather and feathers had hardened, the craftsman pounded the ball to make it round and painted the hard shell white. The far superior flight characteristics of the featherie refined the game. Its prohibitive cost refined the players.

For the next 150 years, golfers played with the featherie. Then in the mid-nineteenth century, golf legend has it that a professor

A rolling history of golf: from bottom, a hand-sewn featherie from the seventeenth century, two molded gutta-percha balls, and later balls made of rubber windings coated with dimpled enamel. (© Sepp Seitz/Woodfin Camp)

at Saint Andrews University received a statue of the Hindu god Vishnu that had been shipped by sea from India. The fragile sculpture had been carefully packed in a rubbery substance called gutta-percha, a substance tapped from certain trees that hardens on contact with the air. Apparently the professor, a fanatical golfer, shaped this alien substance into a small round ball and brought it along on his next golf outing.

At first the gutta-percha ball was an aerodynamic failure, bouncing relentlessly along the ground. But the scholar persisted and noticed than when the club he was using had scored the surface of the ball with tiny nicks, it started to sail. Soon the gutta-percha balls were being mass-produced, with the nicks giving way to dimples made by molds.

A tremendous amount of backspin is imparted to a golf ball on a drive shot, as much as eight thousand revolutions per minute. A golf ball hit to a height of sixty-five feet without backspin will come down after four seconds; a drive shot with backspin that reaches sixty-five feet at the top of its arch will stay in the air for six seconds. These extra two seconds can extend the flight of the ball by as much as eighty feet.

The ball stays in the air longer because of the dimples. The rough surface traps a layer of air that spins with the ball. Since the ball is traveling forward, but with backspin, the layer of air at the bottom of the ball is traveling "against the wind" whereas the top air layer is traveling "with the wind." This means that the air layer at the top of the ball is moving faster than the air layer at the bottom. According to a principle first discovered by Daniel Bernoulli, this difference in air speed creates more air pressure at the bottom of the ball than at the top, and the ball sails, much like an airplane. If the ball were undimpled, it would carry less than one-quarter the distance of a conventional drive.

Professional golfers first balked at the gutta-percha, for the more rugged and less costly new ball threatened their profitable sideline of making the featherie. But the cheaper ball caught on nonetheless, and the game spread throughout imperial England and across the Atlantic to the United States.

Toward the end of the nineteenth century a Cleveland research chemist, Coburn Haskett, began producing balls by winding a rubber thread around a solid rubber core, which was then coated with a hard enamel skin. The modern era of golf technology began.

A recent episode in the conflict between tradition and innovation began in 1975 when Fred Holmstrom, a physicist at San Jose

State University, and Daniel Nepela, a chemist consulting with IBM, decided that a ball could be designed that would minimize the bane of duffers everywhere: the ubiquitous hook and slice.

"We modified existing balls a bit," says Holmstrom, "by slowly filling in some of the dimples and hitting a few drives to see how the changes affected flight characteristics. Then we'd fill in a few more, drive a few, and so on."

In their revolutionary new ball (formally christened the Polara but nicknamed the Happy Nonhooker by jubilant users everywhere), two innovations have been combined. Conventional dimples cover only 50 percent of the ball's surface, and they are restricted to a band around its equator. The golfer places the ball on the tee so that the dimpled band is in the vertical plane. When the ball is hit, the middle band supplies lift, but the smoother sides make the ball less prone to veer off course as it flies. Holmstrom claims that the Polara can reduce a curved shot by as much as 75 percent and still retain all the lift and distance properties of a regular ball.

The United States Golf Association rules specify that a ball must not weigh more than 1.62 ounces and not have a diameter less than 1.68 inches. While the new ball did not violate these restrictions, the USGA became concerned that it would "reduce the skill required to play golf and threaten the integrity of the game." After the introduction of the Polara, the USGA adopted an amendment to the rules that says "a golf ball must be spherical in shape and be designed to have equal aerodynamic properties and equal moments of inertia about any axis through its center." In other words the Happy Nonhooker was illegal.

Holmstrom and Nepela started producing their ball anyway, hoping that popular acceptance would generate enough support to force the USGA to rescind its ban. Unfortunately, people felt reluctant to use a ball not sanctioned by the ruling body and sales fell.

Even without the Polara, a melange of other spherical testaments to an inventor's ingenuity will still be available to the weekend duffer. One of the latest entries is a ball with a pattern of twelve identical pentagons made of four different-sized dimples. The new ball has ten lines around its circumference that do not intersect dimples, lines around which a ball tends to rotate in flight. A conventional ball has three. There are also balls with hexagonal dimples, pie-shaped dimples, and even oversized balls for those who need a bigger target.

Bowling:
The Great Oil Debate

JOHN KIEFER

On Thursday, July 1, 1982, Glenn Allison, a retired professional bowler and member of the American Bowling Congress Hall of Fame, performed an incredible feat—he rolled a 900 series, three 300 games back to back. To do this Allison had to throw thirty-six consecutive strikes. Prior to that Thursday night the highest three-game total ever in league or tournament play was Allie Brandt's record 886 series, rolled in 1939.

After strike number thirty-six, Allison had good reason to believe that the hardest part of his feat was over. But two days later the American Bowling Congress announced that it would not recognize Allison's 900 series. It said the two lanes Allison used showed an abrupt, and illegal, decrease in the amount of conditioning oil along their surfaces. This decrease creates an artificial "wall" that may guide a ball into the pocket and increase the odds for a strike.

Now if the Professional Golfers Association disallowed three consecutive holes-in-one by Arnold Palmer because the grass on the greens had been cut unevenly, that would be news. But this is bowling, where controversies simmer quietly indoors. For his part, Allison still calmly but firmly disagrees. "Their [the ABC] rules never specify what constitutes a sudden decrease," he says. "Other knowledgeable people have reviewed their data and found nothing wrong with those lanes."

On the surface bowling does not seem like a subtle game. After all, what difference could a little oil here or there make to a sixteen-pound bowling ball? Actually, quite a bit.

A bowling lane is three and a half feet wide and sixty feet from the foul line to the center of the number 1 or head pin. A series of thirty-nine interlocking hardwood boards make up the lane and according to the ABC must be level to within four-hundredths of an inch.

To have the best chance of throwing a strike, a right-handed bowler must curve the ball into the 1-3 pocket (the space between the head pin and the pin behind it to the right; it's the 1-2 pocket for left-handed bowlers). Not only is this curve impossible without putting spin on the ball, but a ball without spin will deflect from the optimum path through the rest of the pins. To make the ball spin, a bowler releases the fingers after the thumb.

Each time a bowler releases a ball, the impact creates a pressure of 1,800 pounds per square inch of floor—an average of seventeen times per game and fifteen thousand times per year. To protect the wood from this pounding and to prevent the wood strips from warping, most lanes are coated with urethane. Since the friction between the two slows the ball down, oil is spread on top of the urethane.

Along the first ten to fifteen feet of the lane, which are the most heavily oiled, the ball slides "like tires on ice," says Dave Lumley, manager of ABC's testing and research center, "thereby conserving the ball's spin." As the oil thickness begins to taper off over the next thirty feet, the ball begins to roll. Finally, it hits the "rug" in the last fifteen feet before the head pin, where there is no oil except for what is carried there by previous balls.

Without oil over a long stretch of the lane, friction would allow the spin to take effect too early, making the ball curve prematurely. It would also eliminate the ball's spin before it ever reached the pins. As it is, there's just enough friction those last fifteen feet to ensure that the ball's spin will be translated into the right curve into the pins. For example, suppose a ball is spinning counterclockwise, which happens when most right-handers throw it. For the first part of its journey down the lane it will simply skid on the oil. When it starts to encounter friction, it begins to roll. Because it is also spinning, however, the point at which the ball touches the surface of the lane slowly shifts to the left, and the ball curves left.

The spin can also be affected if a ball's weight is unevenly distributed. A bowling ball is generally made of rubber, polyester, or polyurethane. When finger holes are drilled into it, the ball be-

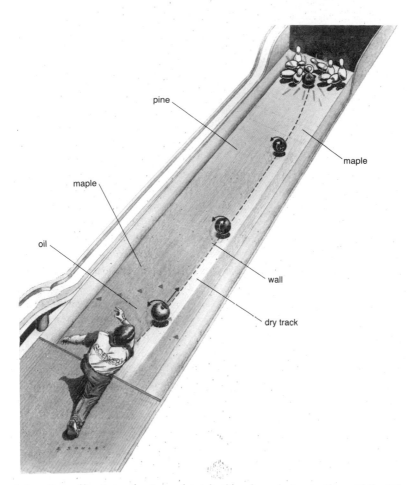

pine

maple

maple

oil

wall

dry track

A bowling alley is made of maple at both ends and pine in the middle. The first fifteen feet of a lane is coated with oil so that the ball slides instead of rolls, thereby conserving its spin. As the oil tapers off, the gripping action of the pine makes the ball begin to roll. When the ball hits the last part of the lane, an oil-free area known as the "rug," the combined spinning and rolling cause the ball to curve into the pocket. Because each ball picks up oil as it rolls, it creates a track of dry wood on the lane. If a ball is thrown along the oil at the edge of this track, above, and begins to drift, the friction of the dry wood makes it hook back and follow the wall of oil down the lane. The American Bowling Congress says Glenn Allison had the advantage of such a wall on the far right side of the lane when he bowled his recordbreaking 900 series. Unscrupulous bowling alley owners sometimes put an extra-thick layer of oil down the middle of their lanes to produce a guiding wall and boost the scores of their clients. (© Robert A. Soulé) comes lighter on top than on the bottom. Bowlers found that when thrown by a right-handed bowler, for example, these bottom-heavy balls tended to curve away from the 1–3 pocket because the surplus weight spent more time on the right side of the ball as it rolled down the lane. To correct this imbalance, manufacturers put heavier material in the top half of the ball. Through trial and error, bowlers eventually discovered that finger holes drilled slightly to the right of this surplus weight made the ball curve more and deflect less because the ball was imbalanced to the left.

"A rolling ball behaves like a gyroscope," says Bill Taylor, coach and bowling guru for more than 30 years. "It likes to maintain its axis of rotation." Thus, when a weight is added to the left side of the ball, it behaves as though a force were pushing down on its left side. But instead of tilting to the left, the spinning ball reacts like the gyroscope and turns left. Because of the variability of grips and size of finger holes, the ABC allows the weight imbalance to range up to three ounces between the top and bottom halves of the ball and one ounce on any side.

All these gyrations have one purpose: to get the ball into the 1–3 pocket with enough spin to carry it on into the 5 and 9 pins. The 5 pin then takes out the 8 pin and you have a textbook strike. In Allison's case the ABC believes the strikes were far from textbook. It argues that the abrupt changes in the amount of oil across the width of the lane formed a wall that helped steer his balls into the pocket. ABC inspectors measure the spread of oil qualitatively by moving their fingers across the lane. Often, as in Allison's case, they will also plot the oil pattern by measuring the distance that a spring-loaded device slides lengthwise down each board.

Bowling alley operators sometimes place a heavy strip of oil down the middle of a lane to give their customers an edge. Such a strip will steer a ball into the pocket because the ball tends to roll along the edge of the oil just as if it were riding a rail. If the ball starts to cross over the oil line, the friction of the drier boards causes it to hook back. Then as it begins to cross back into the oil, the ball slides and keeps riding on the edge.

But the lack of oil on Allison's lanes was limited to just the far right side. "That line to the right side would only make a difference if Glenn was a 'spray bowler,'" says Taylor. "Such a bowler is consistently inconsistent as to where he releases his ball. Glenn bowls 'square'—putting the ball down on the same board each time —and by doing so makes the far right-hand oil line a nonissue, simply because he never releases his ball that far to the right." If what Taylor says is true, someone would have to release the ball farther to the right than Allison and with a lot more spin to take advantage of this oil line. The ABC would probably counter, if it were not officially avoiding comment, that there is no way to prove that Allison was bowling square on the night in question.

Having reached this impasse, the controversy over the 900 series is now headed out of the alleys and into the courts, where Allison is challenging the ABC ruling. Regardless of the outcome, Allison feels that his place in bowling history is secure. "I really would like the series to be officially sanctioned," says Allison, "but for now, it's enough that my fellow bowlers accept it."

Pool-Hall Science

WILLIAM F. ALLMAN

In the early 1800s a French army captain named Mingaud took up the game of billiards at the prison where he was doing time. Sticking a bit of leather on the tip of his all-wood cue stick, he discovered that when he hit a ball off-center—something his predecessors sought to avoid—the extra friction at the tip gave the ball spin and made it do wondrous things. He promptly asked for an extension of his sentence so he could practice.

It was granted. And when he finished his stretch, Mingaud's prowess at the table was unmatched. He then pioneered another billiards tradition: shooting pool for a living. He so dazzled his countrymen with tricks such as curving a ball around a hat that one historian remarked it was "lucky for Mingaud that the statute for sorcery had been repealed."

To a beginner at billiards—whether it's the original game played with three balls and no pockets or the American game of pool—Mingaud's feats may indeed appear otherworldly. Billiard balls appear to carom off each other with an earthy clack as simple and definite as Newton's physics.

Which isn't to say Isaac would have made a great pool player. "It's not like light bouncing off mirrors," says Robert Byrne, a champion billiard player and author of *The Standard Book of Pool and Billiards*. "When it comes to the motion of the balls on the table, the average hustler knows ten times more than a physicist."

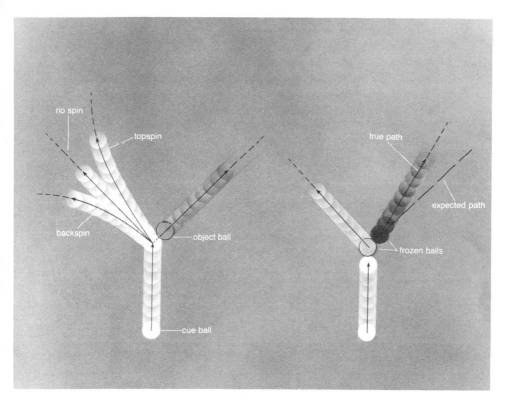

When a cue ball that has no spin hits an object ball, above left, the balls separate at approximately right angles to each other. But because very little spin is transferred to an object ball in a collision, a spinning cue ball will carom at right angles and then curve. A cue ball with topspin, for example, will curve toward an object ball. This is because the cue ball, which before the collision was rolling in a direction parallel to its spin, is suddenly rolling in a direction that is at an angle to its spin. In this case, the spin is angled to the right, and the ball curves right. Similarly, a cue ball with backspin will curve away from an object ball after a collision. Most beginners think that when a cue ball hits two balls that are touching, or "frozen," above right, the furthermost ball will shoot off at right angles from its partner. Because of friction between the two balls, however, the frozen pair behave momentarily as though they were one ball, and the angle between the ball's path is narrower. (© *Flash Fleischer*)

Physics helps, of course. But the difference between a physicist and a billiard player is the difference between theory and practice. Newton played with idealized balls under idealized conditions. A billiard player has to worry about things like the nap of the cloth, the resiliency of the rails, and, thanks to Mingaud, the application, sometimes inadvertent, of spin on the cue ball.

In fact, the only way to make the ball behave as though it doesn't have spin is to give it a little. If the cue ball is hit right on its center it will skid at first, beginning to roll only after the friction of the cloth slows it down. But if the ball is hit a little above center it will get top spin as well as a forward push. A good player can hit the ball so that instead of skidding, it spins at the same rate that it rolls forward, resulting in a shot, called a natural, that does not slide.

When the cue ball hits another ball straight on, it transfers all its momentum to the other ball, called the object ball, which takes off in the same direction the cue ball came in. If it hits off-center, then only part of the cue ball's momentum is transferred and the two balls take off roughly at right angles to each other. One way to gauge where the object ball and cue ball will go after they collide is to imagine a line drawn between their centers at the instant they touch. The object ball will travel along this line, the cue ball perpendicular to it.

This would make things simple except for a phenomenon called throw. In most cases the friction between two colliding balls is very slight. But if the cue ball is hit to the right or left of center—giving it "English"—it will throw the object ball a little in the direction of the spin before sending it on its way. Throw also occurs when two object balls are touching—"frozen," in pool-hall parlance. The friction between the two balls causes one ball to drag another for a short distance before it begins to roll. "When frozen balls are hit," says Byrne, "they behave for an instant as though they are one ball." Thus what might appear to be a sure shot—two balls lined up to the right pocket and touching—may be missed if throw isn't accounted for by hitting the near ball a little to the left of where it looks like it should be hit.

But hitting one ball with another is only part of the game. "It's a cross between chess and golf," says Harry Sims, currently the United States three-cushion billiards champion. "It takes a three-dimensional mind: You have to make the shot, get into a position to make another shot, and make sure that if you miss, you don't leave your opponent with an easy shot." It's the latter two tasks that make Mingaud's work Promethean. Prior to his leatherwork, a cue ball hit another and then rolleth where it listeth. Thanks to English, the cue ball rolleth where the player listeth it.

For example, suppose a pool player wanted to knock in a ball resting near a corner pocket. An easy shot to make—except that the only other ball left to sink lies across the table near the side rail. An expert player can sink the first ball and still set up the next shot by hitting the cue ball below center and hard, giving it backspin. The spinning cue ball skids along the felt until it hits the object ball. When it does, its forward motion will be completely transferred to the other ball, driving the object ball into the pocket. But because the balls are together only an instant, there is little friction between them, and the cue ball continues to spin after the collision. The backward spin makes it roll backward—to an ideal position to sink the next ball. Expert players can give the cue ball so much backspin it travels the length of the table twice.

Similarly, a player can give the cue ball topspin and cause it to roll forward or "follow" after it hits an object ball. English is important for banking off the sides of the table—often necessary for pool and essential for three-cushion billiards, where the cue ball must hit a rail three times each shot. With no English the ball travels a path somewhat like that of light rays reflecting off a mirror, bouncing out at roughly the same angle with which it bounced in. English changes all that. A ball hitting a rail perpendicularly with left English, for example, will rebound out to the left. On oblique rail shots English can make the angle of rebound shallower or deeper and is often used to get to a ball blocked by an opponent's ball.

Sometimes English can be created mid-shot. When a ball with top- or backspin hits an object ball obliquely, it rolls away in a different direction. But the ball's spin, which has not been transferred, continues in the same direction, and the top- or backspin becomes a sidespin. This sidespin curves the cue ball's path. Topspin will make the cue ball curve toward the ball it just hit; backspin makes it curve away.

The ultimate in curving English is the massé shot. The massé was used, no doubt, to make Mingaud's cue ball do its Gallic hat dance. It is also used, rarely, in regular play when the only available shot to an object ball involves curving the cue ball around another ball. To perform a massé shot the cue is held almost vertically, and struck downward on the ball a little off-center. "Essentially, a massé is a miscue," says Byrne. "The ball is squeezed out like a pumpkin seed. It has a slow longitudinal motion, but lots of spin."

An alternative to curving around a ball is jumping over it. Hitting the cue ball very low and hoisting it in the air is illegal, as well as frowned upon by whoever might have to pay for the ripped felt. But a ball hit with enough downward force at about a forty-five-degree angle—a legal shot—will leave the table. The trick, of course, is to make it come down in front of the object ball and not in front of grandma's antique vase.

The massé and jump shots are reserved mainly for show. But the real art of shooting billiards comes in the game of three-cushion, a game that is played all over the world. Billiards uses one cue ball and two object balls. To score a point a player has to hit one of the object balls and at least three rails before hitting the second object ball. Consequently, a single shot can take several seconds as the ball caroms around the table. The shots are often so precise that the nap in the felt, which usually runs lengthwise on the table, has to be taken into consideration because it can make the ball slowly curve.

Five points in a row is considered a good turn, and few masters have scored more than twenty. The complexity of the game gives rise to different styles of play. According to Byrne, Latin American players tend to concentrate on making the shot at hand, Americans stress preventing easy shots for the opponent, and Europeans are fond of setting the balls in favorable positions to string several points in a row.

The Japanese look for help in mathematics. Spaced evenly along the sides of a billiards table, near the cushions, are small spots of mother-of-pearl, sometimes shaped like diamonds. There are seven spots on the long side and three on the short. About fifty years ago someone discovered, through what must have been painstaking trial and error, that if number values are assigned to each diamond and to the position of the cue ball, the number of a spot on the third rail can be subtracted from the cue ball number to produce a number of a spot on the first rail where the cue ball should be aimed. Thus, by using a diamond system, a player wanting his ball to hit the third rail at a particular spot simply does a little fast math, aims, and fires.

Almost. Though the diamond system can improve a player's accuracy, no system completely rids the game of guesswork and and the "feel" of a shot that comes from hours of practice. Sims has some fifty-seven systems he uses to line up shots. Even so, he says, nothing can replace the four hours of practicing he does every day. "Billiards requires a lot of mind power," he says, "but the hardest part of the game is hitting the ball in a straight line."

Why Does a Boomerang Come Back?

DAVID ROBSON

In New York's Central Park a group of spectators and TV camera-men jockey for position while Barnaby Ruhe, holding a boomerang in his right hand, calmly fastens an apple to the top of his head.

He stands quietly for a few seconds, gauging the wind. The launch is abrupt—the boomerang leaves his hand at fifty-five miles per hour with an audible *swish!* Spinning nine revolutions a second, it is visible only as a blur. The boomerang flies outward for a second, enters a broad left turn, and heads back toward Ruhe. As it draws near, the cameramen lean back and Ruhe steps forward. *Whack!* Apple pulp sprays the spectators, two pieces of apple drop from Ruhe's head, and the boomerang tumbles to the ground.

Ruhe performs his William Tell act regularly at boomerang tournaments. "It's a great attention getter," he says, "and it's not as dangerous as it looks—for an expert thrower."

The apparent danger in knocking an apple off one's head is due partly to the boomerang's reputation as a weapon. "There is a popular myth that the boomerang can be used for hunting and that if it misses the game it will return to the thrower," says Ruhe. "But the Australian aborigines used returning boomerangs as playthings—just as we do." Its curved path (which makes it diffi-cult to aim) and light weight (half the weight of a baseball) make it futile for a hunter to throw a boomerang at anything other than a flock of birds.

44

The game-hunting myth no doubt stems from inaccurate reports given by early explorers who confused aboriginal throw sticks with returning boomerangs. The throw stick is a curved piece of wood that is larger and heavier than a boomerang. When thrown, it travels waist-high, in a straight line, for up to two hundred yards. It is a superb weapon, but it does not return.

For many people, just getting the boomerang to return is pleasure enough. But there are always those who will take a pastime and turn it into a sport. For them, there are boomerang tournaments in Europe, Australia, and the United States with competitions such as the maximum-time-aloft event, in which the thrower not only has to keep the boomerang in the air for as long as possible, he must also catch it (the record is 33.2 seconds). In the juggling event a second boomerang is launched while the first is returning. The object is to quickly catch and relaunch it, always keeping one in the air. The world record for juggling is 106 straight catches. Another event is distance throwing. Though boomerangs typically travel outward only about 35 yards, Peter Ruhf of Pennsylvania threw a custom-made boomerang a world-record 125 yards.

The most demanding event is accuracy throwing. The field is marked with four concentric circles several yards apart. The boomerang is thrown from the bull's-eye, and points are awarded for how close the contestant is standing to the center when the boomerang is caught combined with how far away from the thrower the boomerang flies.

In 1981 a team of twelve Americans journeyed to Australia for the first international tournament, and to nearly everyone's surprise won all three matches. The Australian reaction was summarized by the *Sydney Telegraph*: "Yesterday should be forgotten. It should be expunged from memory, deleted from the records. It's enough to make any self-respecting Aussie go bush."

The classic boomerang is L-shaped, with arms of equal length, but excellent returners also have been shaped like pinwheels, with as many as six arms, and like all the letters of the alphabet except B, D, I, O, P, and Q. The principles behind the boomerang's comebacks are well understood. The arms of a boomerang are flat on the bottom and curved on top. As the arms cut through the air, they generate lift, like an airplane wing. Because the boomerang resembles an airplane wing, a novice thrower may try to launch it like one—horizontally, with the flat bottom parallel to the ground. On a "flat" launch, however, the lift generated by the wings pushes the boomerang upward, making it climb steeply.

Attaching a tiny light to one arm of a boomerang, Scottish physicist Mick Hanson photographed its flight to study the curved stick's aerodynamics. (© Mick Hanson)

To launch it correctly, the thrower grasps the boomerang near the tip of either arm and releases the boomerang so that its arms are nearly vertical. Just as an airplane banks its wings to make a turn, the boomerang's "lift," pushes it sideways instead of up.

But the sideways-pushing force alone would merely propel the boomerang to another corner of the field; it is not what makes the boomerang return. An additional turning force is needed, and this is supplied by an enigmatic force called gyroscopic precession. Precession is a phenomenon that affects all spinning objects, be they gyroscopes, bicycle wheels, tops, or planets. If a force is applied to the edge of a spinning disk, for example, the disk will not tilt in the direction of that force, as might be expected. Instead, it will tilt as if a force were applied at a point ninety degrees around the disk from where it was pushed; a spinning disk, when pushed at the twelve o'clock position, will behave as though it were pushed at three o'clock. Thus, to turn a bicycle to the left when riding "no hands," a rider leans to the left, which is the same as applying a force to the top of the front wheel. But the front wheel not only leans left; because of the effect of precession, it also *turns* left.

The "left-turning" force on a boomerang is supplied by the uneven lift generated by its spinning wings. The tips of a boomerang's arms spin with a velocity of about thirty-five miles per hour, forming a "wheel" lying in a near-vertical plane. When an arm passes the top of this disk, it is spinning forward, in the direction of the boomerang's motion. This means that the tip of the arm will move through the air with a velocity that is a combination of the thirty-five-mile-an-hour spin and the boomerang's fifty-five-mile-an-hour velocity, or about ninety miles per hour. When the same arm passes the bottom of the disk, it is spinning backward— opposite the boomerang's forward motion. The arm then moves through the air with a velocity of fifty-five miles per hour minus thirty-five miles per hour, or twenty miles per hour. Because the boomerang arms have much greater speed through the air at the top of the disk than at the bottom, the lift they produce is uneven, with more being generated at the top. This is the same effect as pushing the wheel at twelve o'clock. Because of precession, the boomerang tilts not at the top but at the front edge, causing it to turn left.

Of course, all this may be of little importance to people who love to play catch with themselves. Throwing a boomerang seems to appeal more to the irrational side of the psyche. "There is magic in the experience of throwing a boomerang," says Carl Naylor of Brooklyn, New York. "You've done it many times before, but the rational side of your brain still tells you that something thrown away shouldn't come back. When it does, it's just a bit awesome. I don't get that feeling playing Pac Man."

The Fairy Tale Physics
of Frisbees

MICHAEL GOLD

No one has ever met Harvey J. Kukuk. The only known photograph of him shows a little man in a trench coat standing on a street corner with his back to the camera. Harvey J. Kukuk is the elusive executive director of the International Frisbee Association. He is also a symbol of the uncertainty and mystery that permeate the field of flying plastic disks.

What could be so mysterious about a Frisbee? Even dogs know what to do with them.

Ah, but what do you know about the Hyzer Angle? That's the sideways angle at which most people release a Frisbee in order to ensure level flight. And what of those concentric ridges on the top of the Frisbee, the ones that devotees call the Lines of Headrick?

Few scientists have studied these questions, though Frisbee aficionados have invented an entire language to discuss them. They describe a disk's anatomy in terms of Bernoulli's Plate, the Slope of Shultz, the Bump of Boggio, and Kukuk's Ridge. They have divided its flight into ten phases beginning with the "whelm," passing through the "wax," and on to the "wane," the "waste," and finally the "was." But question an enthusiast about the role of each bump and ridge, or ask what it is that makes a was a was—and you might as well be asking for Harvey J. Kukuk's shoe size.

The first explanation of flying disks came from Fred Morrison, a California building inspector and part-time inventor. In the early

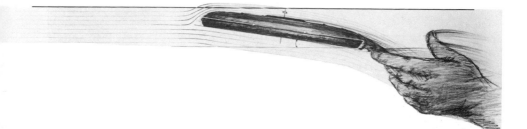

*The gyroscopic stability of a spinning Frisbee maintains the initial angle of attack. As it forces air downward, the disk will receive an equal and opposite force that keeps it aloft. Recent world records: longest throw, five hundred feet; speed, seventy-four miles per hour; time aloft, fifteen and a half seconds. (© M. E. Challinor/*Science 82*)*

1950s Morrison developed the "Pipco Flyer," the "Flyin' Saucer," and the "Pluto Platter," all forerunners of the Frisbee. Although the disks appeared to fly, he told prospective customers, they actually rode along an invisible wire. He demonstrated the miracle at county fairs and sidewalk sales, offering the invisible wire for a penny a foot and throwing in a free disk with every one-hundred-foot purchase.

In 1957 the Wham-O toy company bought Morrison's Pluto Platter, altered the design a few years later, and renamed it Frisbee, for the Frisbie Pie Company of Bridgeport, Connecticut. Legend has it that Frisbie pie tins were the original flying disks, used by students at nearby Yale University in the days before plastic.

When two students from the Massachusetts Institute of Technology embarked on a study of the Frisbee in 1965, they asked Wham-O about the engineering that went into its development. "They told us there wasn't any," recalls Leonard Silver, who now helps design computerized instruments for airplanes. "But they said if we found out why the thing flies that we should let them know."

Despite the lack of research, some scientists will hazard an alternative to the invisible wire theory—at least to explain the basics of Frisbee flight. "The best way to describe it is a combination airplane wing and gyroscope," says one. "If you try to say any more than that, it gets real complicated."

What keeps a Frisbee aloft is probably its platelike shape and its ability to fly forward with its front end tipped up at a slight "angle of attack." Any flat object moving this way will deflect air toward the ground. When the air is forced down, an equal and opposite force is exerted on the disk. That's where the Frisbee gets its lift. It can generate lift even while flying upside down, as long as it maintains an angle of attack. Banking the Frisbee aims its lifting force off to the side, causing the disk to curve.

Unlike an airplane, though, a Frisbee has no flaps and no tail wings to keep it steady. What prevents it from tumbling like a leaf in the wind is its spin—angular momentum, to be exact. Any rotating object, a football or a bicycle wheel, for example, has a strong tendency to maintain its orientation in space. The faster it spins and the more its weight is distributed in the outer edges—where most of a Frisbee's weight is—the more stubbornly it holds its orientation.

The concept of a gyroscopic airplane wing explains basic Frisbee flight so nicely, it seems a shame to bring up the Hyzer Angle. Almost every beginning Frisbee thrower, and a good number of veterans, release the disk with the side away from the hand tipped noticeably downward. What's shocking is that this angled release achieves a straight, level flight. Somehow, within the first few feet of its journey, a Frisbee thrown this way will kick up and level off. What happens to the gyroscopic stability? Why doesn't that lowered side remain low and the disk follow a curved path?

"It's possible that there is some critical Hyzer Angle below which the disk always rights itself and beyond which it always executes a banked curve," says Jay Shelton, a former Frisbee distance champion with a doctorate in physics. "But I'm certainly not going to try to figure the physics of that question."

Others less bound by the shackles of science are more willing to theorize. "A novice needs to use a Hyzer Angle because his fingers don't release cleanly," says Dan Roddick of the International Frisbee Association. "I think it has something to do with yanking."

We have entered the realm in which scientists fear to speculate and Frisbee aficionados cannot make themselves understood. What better place to consider the Lines of Headrick, the circular ridges near the Frisbee's edge. The addition of these lines was one of the few major design changes Wham-O made on Morrison's Pluto

Platter with a specific goal in mind. Names for their inventor, Ed Headrick, Wham-O's former vice-president and director of research, the patented lines were to simulate devices on airplane wings known as vortex generators.

Under some conditions the air flowing over a wing will suddenly break away from it, leaving a low-pressure wake on its trailing edge. This creates a suction force that fights the forward progress of the plane. Vortex generators are small tabs that stir up turbulence, mixing the separated air stream back into the low-pressure region.

The Lines of Headrick might play a similar role except that, like the rest of the Frisbee, they spin some three hundred times a minute. The resulting airflow is probably so complex that it is impossible to predict the effect. "What relationship these lines have to real vortex generators is anybody's guess," says Mel Zisfein, former deputy director of the Smithsonian's Air and Space Museum.

Headrick says he verified that disks bearing his lines suffered less drag in wind tunnel tests he conducted in the mid-1960s. The only independent study, done by a group of Canadian students, found that Frisbees flew better without them.

Perhaps it is best not to delve too deeply into these secrets. As E. B. White once said of humor, it can be dissected like a frog, "but the thing dies in the process and the innards are discouraging to any but the pure scientific mind." Maybe the Lines of Headrick, Boggio's Bump, and the Hyzer Angle should remain unexplained.

"The Frisbee may be a subtle and sophisticated aeronautical device," says Zisfein, "but it's also fun just to throw."

Kukuk couldn't have said it better.

Part II

Soul of a Running Shoe

ERIC PERLMAN

Why run in shoes at all? The naked foot was good enough for antelope-chasing prehistoric hunters, ancient Greek Olympians, and even a few modern runners such as Britain's Bruce Tulloh, who set distance records in the 1950s and '60s while racing unshod. But the world is running short of open savannas, and what were once dirt roadways are now coated with the bane of many an Olympian and weekend athlete alike: pavement. "We weren't designed to run on it," says Peter Cavanagh, who has spent a good deal of his career as a biomechanical engineer studying how runners run and how the hazards of running take their toll.

Rocks, broken glass, and abrasion are just a few of the obstacles confronting a runner. Running only fifteen miles a week—and many competitive runners do ten times that much—means that the foot strikes the ground about a million times a year. That's a lot of pavement hitting the foot, and if shoes don't absorb that shock, it goes to the ankles, shins, and knees.

Roughly one-third of all marathon runners suffer injured knees, inflamed tendons, shin splints, and blisters every year. And as people are usually willing to spend money to avoid pain, sales of shoes claiming to protect the foot have boomed along with the sport's popularity. Some manufacturers are runners themselves who are trying to assure their comrades more comfort and fewer

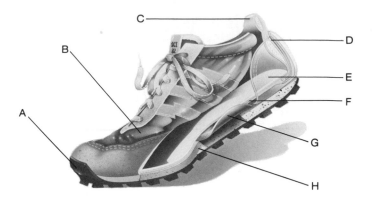

A—OUTSOLE   It can be patterned with ribs, waffles, or ridges. A deep tread pattern like a waffle offers better traction on rough terrain but wears faster on pavement. Rippled soles have the most surface area, so they wear down least. Some waffle soles have extra rubber in the heel, where most runners hit the ground first.

B—UPPER   Usually made of nylon, it holds the sole of the shoe onto the foot. Its seams can be a source of blisters, but the biggest challenge to manufacturers is finding a fabric that allows the shoe to breathe while repelling water. Despite some claims, no material does this. Restricted air flow of some shoes can raise temperatures inside by more than fifteen degrees Fahrenheit.

C—ACHILLES TENDON PROTECTOR   This tab of padded vinyl is designed to ease pressure on the Achilles tendon, which runs along the back of the heel, but it can actually irritate the tendons of some runners.

D—HEEL COUNTER   This semicircle of plastic or fiberboard stiffener is sand-wiched in the heel. Most runners strike the ground on the outside of the heel and roll the foot toward the center. The heel counter helps prevent the foot from sliding to the inside during the roll.

E—SOCKLINER   Lying directly under the foot, this thin layer of foam rubber, crushed velour, or terry cloth helps to absorb sweat and shock and to prevent blisters.

F—INSOLE BOARD   A flat piece of cellulose fiber, it serves as the skeleton to which the other elements of the shoe are bonded. The board spends a good deal of time bathed in sweat and is therefore often treated with chemicals to combat bacteria and fungus.

G—ARCH COOKIE   This wedge of foam is intended to lend some support to the arch of the foot. It does little more than give the shoe a snug fit.

H—MIDSOLE   Made from foam containing tiny air bubbles, it absorbs the impact of the foot hitting the ground. The air cells, however, can collapse; some shoes lose much of their shock-absorbing properties after a hundred miles. The wedge, also a cushion, raises the heel to ease tension on the Achilles tendon.

injuries. Others, drawn by the $25 to $100 prices, are in a run for the money.

Like most shoes, running shoes are generally made with an outsole, a midsole, and an upper. But running shoes employ a few features that are not standard equipment on a pair of sneakers. The midsole, responsible for shock absorption on most running shoes, is the designers' favorite playground. One manufacturer has hollowed out chambers of foam and filled them with inert gas. Others have tried air pockets that inflate through a valve at the side of the shoe, canals cut out of the heel region that give on impact, or tiny holes beneath the ball of the foot that increase flexibility.

A new development in outsole design is the incorporation of removable plugs of hard rubber beneath the heel, the ball of the foot, and other areas that get the most wear. When they are worn down, a runner simply pops in new ones. Unfortunately, the plugs protrude beyond the sole and thus concentrate the force of the body's impact on landing on a smaller area of the foot. Meanwhile, a Japanese manufacturer has introduced an insole with replaceable inserts so a runner can adjust the amount of shock absorption and stability in the shoe.

Some shoe manufacturers have tried to correct what they see as mistakes in runners' bodies. The Varus Wedge, for example, was a tapered midsole that lifted the inside of the foot, causing it to slope toward the outside though the sole itself remains flat. This was an attempt to counteract the natural tendency of the runner's foot to land on the outside of the heel and then roll toward the inside as it rocks forward for the next step. Some runners do this movement excessively, causing increased stress on the knees and ankles. Many runners still swear by the Varus Wedge as an improvement in mechanical efficiency and injury reduction. Others swear at it as the worst knee killer since the rack. According to orthopedist Roger Mann, "Less than 15 percent of all runners can benefit from the Varus Wedge. The other 85 percent are either being hurt or not helped."

Another development adopted by most shoe manufacturers is the flared heel, designed to keep feet from rolling from side to side as a protection from ankle sprains. It works very well. Following the "more is better" philosophy, one manufacturer added a full inch of heel flare, about 50 percent more than average. But runners across the country angrily returned the shoes, complaining of increased knee strain and severe pain. The excess flare of the heel apparently caused the foot to rotate inward too fast.

The most expensive addition to running equipment is the orthotic support, a custom-fitted shoe insert set between the bottom of the foot and the insole that changes the way the foot and leg are oriented when they hit the ground. Custom orthotics cost between $200 and $300. Though some doctors feel they can correct specific running problems, Mann says they are "overpriced and overused. All they do is provide extra cushioning at the arch, and most people don't even need that."

But the ultimate worth of a shoe is not dependent on the gizmos built into it but on how it fits the way a runner runs. The heavyset cannonball who shakes the ground with every step needs a lot of shock absorption. The human gazelle may need less padding but requires more support and control to prevent turned ankles. The truth is that people who run only a few miles a week can wear just about anything—wing tips to flip-flops—as long as the shoes are comfortable.

For the more serious runner, one of the biggest problems occurs at the shoe store. It is hard to distinguish the useful new development from the useless gimmick. Until 1981 Cavanagh, professor of biomechanics at Pennsylvania State University, conducted an independent running shoe testing service. He checked each year's crop of running shoes for flexibility, shoe weight, sole wear, traction, water permeability, support, and shock absorption.

Cavanagh found, for example, that most shoes last only about five hundred miles before the wear on the sole and the loss of cushioning decreases their usefulness. He also found that a thicker midsole of wedge on some shoes did not always mean more shock absorption and that a good sockliner, the inside surface that cradles the foot, can add as much as 20 percent more cushioning.

The survey was a popular guide, and because a high score meant big sales at shoe stores, it pushed manufacturers to improve running shoes as well. Unfortunately, Cavanagh's survey was discontinued in 1982 and runners once again have to rely on word-of-mouth advice.

According to both Cavanagh and Mann, there is still plenty of room for development in running shoe technology. Many running shoes for women, for example, are simply smaller men's shoes with new packaging and don't take into account the different biomechanical needs of women runners; for example, a woman's heel is proportionately thinner than a man's. For future runners Cavanagh envisions a shoe that comes to the shoe store in pieces so that the buyer can customize the various elements to fit his needs. "It would be like walking into a tailor's shop. You would get your measurements taken and have a shoe custom-built."

Tuning the Track

ANTHONY CHASE

For years the members of Harvard University's track team had been working out on a packed cinder oval in a drafty shed that also housed the baseball team's winter batting cage. The runners were protected from stray fastballs and line drives by a shabby net that hung from the ceiling, but there was no way to protect their legs from stress injuries brought on by training hundreds of miles on such an unforgiving surface.

When Harvard's coach was finally able to convince the planning office to build a new track and field facility in 1976, he asked two of the university's biomechanical engineers, Thomas McMahon and Peter Greene, to design a track surface that would be gentler on his runners' legs, knees, shins, and feet. "We told him that a safe and comfortable surface, one that was sufficiently springy or 'compliant,' would be unlikely to produce fast times," says McMahon. "He just told us to do the best we could. Our primary consideration was to be the safety of the athlete."

At that time, no one really knew what effects various surfaces had on running speed and comfort. Indoor tracks were built of cinders, plywood, and even concrete beds covered with a layer of rubber or polyurethane. In most cases assumptions about the effect of the surface on running times and injuries were based on the intuition that the hardest tracks are the fastest and that concessions to injury-prone runners detract from the "speed" of the track.

"Most people figured that since the coaches were the ones ordering the tracks and they are primarily interested in extracting the fastest possible performances from their teams, they would be more willing to accept the risk of training injuries brought on by pounding around on the harder surfaces," McMahon says.

McMahon and Greene began their research by examining the act of running itself. Because running is essentially a succession of bounds, they were able to develop a mathematical expression for the "stiffness" of the leg's muscles and skeleton that make up the human "spring." By feeding this information into a computer, the researchers determined the optimum stiffness for a track.

McMahon and Greene then designed a mechanical model of the running leg, one that was realistic yet simple enough to be mathematically manageable. A gear system represented the leg muscles responding to neural signals from the brain, a coiled spring illustrated the stiffness of the muscle, and a shock-absorbing mechanism represented the interwoven system of muscles that dampen the body's impact with each stride. At first the researchers had not included the nerves and muscles of the reflex system in their calculations, but "when we threw a shock-absorbing term into our equation," says Greene, "bingo! The whole thing came together."

Once this evaluation of the stiffness of a runner's muscles had been included in their calculation, the computer spat out a startling projection: An ideal surface was about twenty times springier than most current designs. In order to evaluate these unexpected results, the researchers set up a series of test tracks ranging in stiffness from unforgiving concrete to the sponginess of high jump landing pillows. They also used a set of plywood tracks with adjustable spring. Twenty runners were hired to race up and down the basement hallways. A force-measuring device beneath the plywood panels recorded the impact of each runner's collisions with the floor, and a high-speed camera filmed the springs at two hundred frames per second. A computer reduced the full-figure images to a series of connected line drawings.

The computer diagrams indicated clearly that though the step length—that is, the distance a runner travels while one foot is on the ground—increases on a very soft surface, the time the foot stays in contact with the track also increases considerably. The net effect is to slow the runner. The results also show that contrary to old assumptions, the hardest surfaces are not the fastest. When the surface is absolutely unresilient, the muscles and ligaments of the runner's legs are forced to perform all the shock-absorbing functions, which in addition to being the major causes of injury, is more time-consuming than if the impact is shared equally by the track and the leg.

McMahon and Greene also found that a hard surface returns less energy to a runner than a more compliant one. At every step in a race a runner stores energy in the surface of the track directly beneath his feet, much the way a pole vaulter stores energy in the pole as it is planted and bends. Like the pole, which returns the energy to a vaulter and sends him over the bar, a springy track restores the runner's investment of energy in each stride. Thus, if the surface were accurately "tuned" to complement the stiffness in the average runner's legs, it could not only act as a shock absorber but also as a spring that sends him forward into each stride with an additional amount of energy, thereby increasing his speed.

The results of all the vigorous sprinting and scrupulous measurements confirmed the test model of a track that is about twice as stiff as a runner's leg. This model was then used in the most important experiment of all: the construction of Harvard's new track.

When the track had been in use for a couple of seasons, it was possible to compare the times team members turned in at Harvard with those recorded at away meets. The theoretical predictions made by McMahon and Greene were confirmed: Not only were

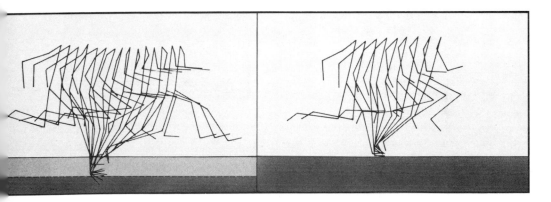

These computer-produced sequences show that while step length is longer on a very soft surface, left, than on a very hard surface, right, the length of time a runner's foot is on the ground also increases, resulting in slower running speeds. (© Tom McMahon)

injuries cut in half, but the average runner's speed was 2.91 percent better at home than on the road, even though two of the away meets were championships. In the previous year the team ran an average .26 percent slower at home.

Similar tracks have been built at Madison Square Garden and Meadowlands Sports Arena in New Jersey. Since their installation, thirty-four world, American, or indoor track records have been broken there.

Because lap lengths of indoor tracks differ, world records are recognized only on outdoor surfaces. McMahon and Greene are confident that when their track is built outdoors some records will be shattered. "There's no reason why the world record for the mile shouldn't be bettered by five to seven seconds," McMahon says.

Unfortunately, their tracks cost nearly twice as much as conventional designs. A pilot project at Loughborough University in England, home of world-class miler Sebastian Coe, ran out of money before it could be built. At present no other tuned tracks are under construction. Apparently, the world will have to wait until someone decides that five seconds off the mile is worth the money.

Artificial Turf: Is the Grass Greener?

ROGER RAPOPORT

In 1979, bowing to player complaints about injuries on synthetic turf, the city of San Francisco replaced Candlestick Park's half-million-dollar, eight-year-old artificial turf with good old-fashioned grass. While many 49er gridiron stars were delighted to switch back to the real thing, it soon drew some negative press of its own. Poor drainage, inadequate rooting, and heavy rainfall turned Candlestick into the bane of the NFL. By the end of their 1981 championship season, the San Franciscans' home field had become known colloquially as Candleslop Park.

The mud at Candlestick is only a recent chapter in the twenty-year-old rivalry between artificial and grass playing surfaces. Pioneered in the mid-1960s by the Monsanto Corporation, nylon turf now covers more than three hundred playing fields around the world. Yet with more than thirty million square feet of artificial turf on the ground, players continue squabbling over it. Some, like St. Louis Cardinals offensive lineman Dan Dierdorf, love it: "Give me AstroTurf anytime. When I dig in on AstroTurf, I get positive traction. I can concentrate on the play, not my footing." Others share the much quoted view of former baseball star Richie Allen: "If my horse can't eat it, I don't want to play on it."

Athletes who advocate restoring the natural grass blame artificial turf for tendinitis, abrasions, shin splints, and broken bones. Others insist that AstroTurf makes the ball bounce higher or throws their passing off. And some complain that hot days are even hotter on a synthetic field than on grass.

AstroTurf is made of nylon fibers roughly five-eighths of an inch long stitched into a tightly woven mat made from the same polyester cord used to produce automobile tires. It is then laid over a five-eighths-inch-thick foam pad for cushioning and glued to an asphalt base. The result is a field that forces players to adjust their game to some new steps and bounces. On grass, says podiatrist Joseph Dollar, an athlete "will use his great toe to dig downward and grip in a claw-type action. . . . However, when running on artificial turf, he is not allowed this gripping action and must make an adjustment in the movement of the first phalange." The first phalange is part of the big toe, and while moving it outward improves the grip, it also puts stress on the muscles around it, creating the painful inflammation called Turf Toe and a related ailment, Turf Joint.

Another problem with AstroTurf is that nylon tenaciously resists tearing. While this makes synthetic fields more durable, it also means that superficial skin wounds known as rug burns are more common, forcing players to wear extra padding. Football and baseball players are most concerned, however, that artificial turf might put them on the disabled list. Football knee injuries, for example, commonly happen when a player is tackled at a moment when the foot is planted on the turf. Some athletes argue that AstroTurf's better traction makes the knees more susceptible to serious damage.

The National Football League Players Association repeatedly tried to have artificial turf declared a "hazardous substance" and have it removed from stadiums. But the Consumers Products Safety Commission concluded in 1973 that the evidence did not show that persons playing on fields covered with AstroTurf incurred a greater risk. Unpersuaded by this view, some coaches like Dan Reeves of the Denver Broncos (who lost all six of their road games played on AstroTurf during the 1981 and 1982 seasons) believes his players are safer working out on grass and only uses an artificial practice field so rookies can get the feel of the stuff.

It's not uncommon for traveling players to carry three or four different types of shoes. According to Bronco spokesperson Erwin Boettcher, "On AstroTurf some will use soccer-type shoes with eight half-inch plastic cleats. Others prefer basketball-type hightops with fifty small plastic cleats. And a few take their cue from retired Pittsburgh star Mean Joe Greene and just wear tennis shoes." A 1973 study by Joseph Torg, an orthopedist and director of the University of Pennsylvania's Sports Medicine Center, indicated that a soccer-type molded sole shoe with fifteen half-inch cleats could safely replace any mixed bag of footwear on both grass and AstroTurf.

Monsanto, the manufacturer of AstroTurf, believes that most other player gripes about artificial fields can be resolved through similar technological fixes. Reflecting on the heat problem that persuaded Miami's Orange Bowl to dump synthetics for grass in 1977, AstroTurf spokesperson Don Berns says that if player comfort becomes an issue in hot weather, the easiest remedy is simply to water the grass. The University of Washington also wets its practice fields to make them more slippery and thus safer. The players didn't like playing on the wet surface at first, but after a while they didn't want to play on a dry field, and the university began to wet the AstroTurf for its games.

One problem that water won't fix is the timing adjustments that some players accustomed to grass fields must make when they switch to nylon. Among the early casualties was Archie Manning, who threw four intercepted passes on the first afternoon he quarterbacked Ole Miss across an artificial gridiron. Game films showed that he had underthrown his receiver by about a yard and a half each time. It appeared that the receivers were running slightly faster, which Manning did not anticipate. Monsanto's Berns offers an additional explanation: "To promote drainage the AstroTurf field is designed with a crown down the middle and a three-percent grade sloping down to the sidelines. This feature means that a player sprinting to the sideline is going downhill. In addition the receiver standing at the sidelines is about a foot below the level of the quarterback at midfield."

Players also must remember that synthetic fields do not all play the same. Ultraviolet radiation, smog, dust, and traffic all cause AstroTurf to deteriorate. Researchers working at the West Virginia University Medical Center found that the time it took for a shot put to come to rest on synthetic turf that had been in use for five years was 7.6 seconds, compared with 11.4 seconds on new AstroTurf. The faster stopping time means that with age the AstroTurf becomes less springy and therefore less shock absorbent, which is why it generates more complaints from players about tendinitis and shin splints.

A new synthetic field can be installed for $600,000 to $700,000 (about double the cost of a grass stadium) and includes a few new innovations such as pencil-thin holes in the mat that allow water to drain through and something Monsanto calls textured turf. The fibers of early versions of AstroTurf often became matted down during play, giving the field a varied "grain" that changed a player's speed depending on whether he was running with the grain or against it. The blades in the new turf, says Berns, are curly instead of straight so they resist matting.

A natural alternative is Prescription Athletic Turf (called PAT), which is grass laid over a network of perforated pipes that suck a wet field dry. Both the Orange Bowl and Denver's Mile High Stadium have such a turf. But a torrential rain before the 1982 AFC championship brought complaints from the New York Jets (who lost the game in the mud) that the Orange Bowl field, which was not covered by a tarp prior to the game, was improperly cared for. The National Football League has since adopted a rule requiring all grass fields to have tarps.

AstroTurf is certainly quagmire-proof, but many athletes remain suspicious of it. Berns responds, "There are still a lot of complaints from athletes, particularly older athletes. They don't like the idea of playing on something different. They have a tendency to use it as an excuse for their own problems. These players are just getting older, and they blame the artificial grass."

The Tennis Racquet

JAMES RANDALL

Ten years ago, when tennis togs were as white as the tennis balls and tennis racquets were thumpingly wooden, the sport moved at the languid pace of the master craftsmen who laminated racquets of selected hardwoods for a handful of established companies. More recently, along with the radical change in the way tennis players dress, play, and behave, there's been an equally dramatic change in their racquets.

When John McEnroe defeated Chris Lewis in the 1983 finals at Wimbledon, the young Lewis wasn't the only up and coming phenomenon on the court. Both he and McEnroe played with "mid-sized" racquets—racquets whose faces are larger than those of conventional models. In fact, the "conventional" model isn't the convention any more. According to the director of racquet production at Le Coq Sportif, models with larger faces account for more than 60 percent of racquet sales, and Le Coq Sportif doesn't even make a standard model. The army of pros who wore down the grass at Wimbledon not only played with a bewildering assortment of shapes, their racquets were made of materials that only a few years ago were more likely to be seen in the space program than on the tennis court.

Much of this change has been stimulated by savvy marketing designed to woo the consumer, in addition to a real effort to improve the product. But it is also due to a hazy definition of tennis' most important piece of equipment. While football, baseball, and basketball rules are precise about the size, shape, and weight of the gear, the U.S. Tennis Association merely states that a tennis racquet is "the implement used to strike the ball."

Little wonder then that a German fellow a few years back invented something called the spaghetti racquet—a standard frame whose double strings, fitted with plastic tubes and other doodads, made the racquet look like a macrame artist's nightmare.

The real nightmare was on the court, where a ball hit with the bizarre racquet had so much topspin that it often caromed right out of play. Suddenly, unsung players were trouncing top pros, and when Ilie Nastase used a spaghetti racquet on Argentine pro Guillermo Vilas, Vilas stormed off the court. In the furor the International Tennis Federation hurriedly devised a new rule of tennis: While allowing racquets to be "of any material, weight, size, or shape," it specified that attachments to the racquet must not alter the flight of the ball and that the strings must be evenly spaced.

The new rule, of course, left the designers with plenty of racquet to play with. Today's racquets are made with materials like graphite, boron, magnesium, titanium, plastic, and fiber glass. Even some traditional wood racquets have become sleek hybrids of wood mixed with graphite and other nonwood materials.

Graphite, the trendiest of the new materials, is perfect for tennis. It is lighter, tougher, and ten times stiffer than wood. Stiffness produces power and helps reduce the twisting of the racquet head on off-center shots that causes wayward, punchless returns.

But scientists also found that too much stiffness resulted in lack of control and caused jarring vibrations that produced sore arms. So they enhanced the shock absorption of the racquet by designing hollow graphite frames filled with plastic foam. Control was improved by designing a bigger sweet spot, the precise area on the racquet face that delivers the most powerful shot with the least amount of vibration. The sweet spot on the average racquet lies about two inches from the base of the head. Unfortunately, most weekend tennis buffs hit the ball in the center or top of the racquet.

Howard Head, who popularized modern metal skis, discovered that a racquet with a bigger face had a bigger sweet spot. Just how much bigger turned out to be something of a surprise: Increasing face size by 20 percent resulted in a 300-percent increase in the size of the sweet spot. It was Head's racquet, which now claims 25 percent of the market, that set off the revolution in tennis equipment.

Before his racquet went to court, however, a good deal of testing was done. Head's researchers used high-speed film to follow tennis balls as they hit racquets. In this way they could measure the coefficient of restitution (how much velocity the ball retains after it strikes the racquet face) and the angle of return (the angle from which a ball rebounds from the racquet face).

What the scientists were looking for was a racquet with a high coefficient of restitution near its center, which would result in an ideal sweet spot. They also wanted a small angle of return, for then the racquet would twist so little on impact that the ball would be rebounding off the strings in an almost straight line, giving the player more control.

Fischer, an Austrian ski manufacturer, also markets a new fiber glass graphite racquet with a roughly diamond-shaped head. Fischer's researchers found that this unusual racquet accomplished two things: The jarring vibrations that occur when a ball strikes the strings were shifted to the end of the handle, where the natural damping of the hand takes over; and an elliptical sweet spot extending toward the top of the racquet face was created.

A composite racquet introduced recently by Le Coq may make the most of the new materials. The racquet head is attached to the handle by three prongs. The center prong, made of fiber glass, absorbs vibrations. The outer prongs, graphite–fiber glass composites, provide rigidity.

New materials have also fostered new ways to make frames. For example, one company impregnates graphite fibers with epoxy and then wraps the fibers around a polyurethane foam core. The number and direction of the windings determine the playing characteristics of the racquet, much as the plies in an automobile tire help determine the way a car will ride.

All racquet makers human-test their products before marketing them. Rossignol, a French firm noted for its skis, electrically wires tennis players to racquets. A solid-state vibration sensor worn on their chests and backs automatically records every twitch in the racquet at the moment of impact.

Of course, a good, better, or best racquet can't compensate for the built-in handicaps of the person using it. That a really good player will beat a so-so player, regardless of the racquet that the latter happens to be using, was humorously demonstrated by Bobby Riggs. He once beat an opponent when using a broom.

The latest attempt at an on-court advantage is a racquet designed by engineer John F. Bennett of Dynamics Operational. The lower end of the racquet is bent at a nineteen-degree angle. When the racquet is swung level the wrist is slightly cocked, which, according to Bennett, is a more stable position that reduces the risks of tennis elbow.

The newest entry in the tennis racquet game has a handle bent at a nineteen-degree angle. Its inventor claims the kink helps stabilize the wrist and reduce elbow injuries. (© Fil Hunter)

A Mountaineer's Best Friends

ERIC PERLMAN

1934—A mountain climber grips the steep rock with waning strength. His hobnail boots scrape at the granite. He screams for help to his partner, who tends the frayed hemp rope from around a rocky corner. The screams go unheard in the high wind. Desperately, the climber pulls out a soft iron piton, stuffs it into a crack, and pounds it with his alpine hammer. He attaches a quarter-pound iron snap link, called a carabiner, into the eye of the piton and clips his rope into the carabiner.

Safe, at last, he tells himself. Then he slips, dislodges a piton, snaps the rope, and tumbles to the ground one thousand feet below. In 1934 climbing was not a very popular sport.

1984—A mountain climber steps delicately onto a rock edge the thickness of a dime. His high-friction shoes do not slip. He whispers into his walkie-talkie-wired climbing rope, and, despite the howling wind, calmly informs his partner that the next section of rock looks severe and deserves some protective hardware. He grabs a spring-loaded camming device, called a Friend, and wedges it with one hand into a two-inch-wide crack, then clips his rope into it with a one-and-a-half-ounce chrome-molybdenum alloy carabiner.

His finger strength fails, and he falls. No problem. The Friend and carabiner can withstand more than 3,500 pounds of impact with gentle, shock-absorbent strength.

Equipment for the modern rock climber: high-friction shoes, braided filament rope, and Friends. The gear is expensive (Friends cost between $35 and $45 apiece) but can save weight and lives. (© James Balog)

Advances in equipment have improved climbing capability and safety so greatly that despite nerve-racking exposure to falling rocks and weather and the ever present danger of falling, climbing has become one of the fastest-growing sports in America. In 1958, for example, 388 climbers reached the 14,410-foot summit of Mt. Rainier in Washington. Last year almost 4,000 climbers made it to the top.

The most radical advance in climbing hardware since the nylon rope is the Friend, developed by Ray Jardine, a master rock climber and former space flight mechanic in Colorado. Jardine was frustrated with the awkwardness, weight, and marginal safety of conventional climbing hardware, which consisted of iron pitons that were driven into cracks with a hammer (often a two-handed operation), and nuts, wedge-shaped chunks of metal that were slotted into cracks like nickels into a vending machine. "Finding the right nut to fit a crack can be time-consuming," Jardine says, "especially if you're hanging on for dear life."

"We were looking for material that produced a lot of friction between metal and rock but was also incompressible," Jardine says.

He chose an aluminum alloy that held against rock ten times better than iron or steel.

Jardine worked out the best shape for the comma-shaped cams so that they would grip and hold with a constant force regardless of their orientation. Each Friend has four cams that are independently suspended so they can flare out to adjust to widely different cracks. Unlike the nuts, which can only be used on cracks that widen and then narrow, the Friends can hold in cracks that open out as much as thirty degrees. And Friends come out of a crack as easily as they go in—a "trigger" on the stem pulls in the cams to their narrowest setting, releasing their grip.

The heart of a climber's safety system is the rope. It must be strong enough to hold a 180-pound climber for a fall of more than fifty feet, yet it must stretch and absorb the shock of impact. Thin steel cable is lighter, stronger, and more resistant to cutting than nylon, but it does not stretch at all. The poor climber whose fall was stopped by steel cable would probably snap his spine. On the other hand, a rubber rope would be virtually shock free, but the

A Friend's grip.
(© James Balog)

stretch would be so great that a falling climber would probably slam into a ledge or other rock outcrop even while the rubber rope was saving him from the shock of the fall.

Climbing ropes were made first from the hemp plant, then natural silk, then twisted nylon. Modern ropes are made of Perlon, a synthetic material resembling nylon that combines strength with elasticity. The rope is constructed in two parts—an exterior sheath, woven to resist cuts and abrasions, and an inner core made of thousands of braided filaments that run the length of the rope.

Edelrid of West Germany, the world's leading climbing rope manufacturer, weaves its rope cores with 50,400 threads, each with a diameter of 1/100,000 of an inch. There are more than 2,500 miles of Perlon thread in a standard 165-foot climbing rope with a diameter of a little less than half an inch. This microscopic distribution of impact is the key to the climbing rope's lifesaving strength and resiliency.

The most exotic development in rope technology is the "talking rope," which has a built-in, battery-powered intercom. The communication line is coiled through the interior of the rope and stretches out with the impact of a fall. Talking ropes are especially useful in high winds, inside rock chimneys, and next to thundering waterfalls.

Modern rock climbing shoes look and perform more like ballet slippers than mountain boots. They are tight fitting for extra leverage and control. The toes of the shoes are narrow and tapered for slotting into inch-wide cracks. The sole is smooth and pliable to resist slipping on climbing surfaces that may consist of nothing more than a few hundredths of an inch of crystalline bumps on a slab of granite.

The composition of the sole is the key to modern rock climbing technique. While most shoe manufacturers spend money researching ways to harden the rubber and increase sole longevity, climbing shoemakers have refined the science of softening the rubber. By juggling the recipe for compounding rubber, they've made it almost sticky.

A few European rubber makers dominate the market and zealously guard their high-friction recipes. The turnover is rapid—sticky-soled climbing shoes wear out after about a month of daily use—and the profit margin huge. An average pair retails for $80 to $100. Not that there is much of an alternative. To climb the severe routes that were unthinkable thirty years ago but are well traveled now, even the best climbers could not get off the ground without their high-friction shoes.

Using gymnastic chalk for a dry grip, a climber scales a rocky face in Colorado. (© James Balog)

Cross-Country Skiing: The Hows of Wax

ANTHONY CHASE

The U.S. Cross-Country Ski Team ran the entire length of Vermont, a distance of more than 230 miles, in nine straight days to train for the 1972 Winter Games in Soporo, Japan. The 1984 team still does a lot of roadwork, but now much of it is on "roller-skis," short skis with wheels on which they trek several thousand miles a year. Such dedication to distance is necessary for races than can last as long as two and a half hours and cover thirty-two miles, winding through forests, up steep hills, and across broad fields. But the winner among dozens of world-class skiers is not always the one in the best shape. For in competitive cross-country skiing, sometimes it's not who's on top of the ski that counts, but what's on the bottom.

The undersurface of a cross-country ski is coated with several layers of wax, and if the consistency of the wax isn't just right, a competitor is probably better off walking. Racers have been known to step out of a race to rewax their skis, sacrificing precious seconds if they feel that they are sticking too much or sliding too easily.

"Trying to race with the wrong wax sometimes feels like you're glued to the snow. It can be a nightmare," says John Caldwell, former coach of the U.S. cross-country team. "All the conditioning in the world isn't worth a damn if you've waxed wrong."

A downhill racer requires only that his skis slide as easily as possible over the snow. His speed is the result of the force of gravity; he literally "falls" down the hill. To travel across flat ground and over hills, however, cross-country skiers must be able to "run" on the snow as well as glide. To do this, the skier pushes off with one ski while shifting his body weight to glide on the other. Ski poles are used to help supply balance and rhythm.

Four thousand years ago nomadic hunters in Siberia wore a short, wide ski on one foot and a long, narrow ski on the other. Caribou skins were fixed to the bottom of the short ski to provide friction for pushing off; the hunter would glide on the long ski. A version of this device using strips of synthetic fur is available to the recreational skier today.

"It's one of those sports that is very simple in theory and enormously complicated in practice, especially at the level of Olympic competition," says Morten Gaarud, a wax expert at Swix Sports. "The consistency and structure of snow crystals varies at different temperatures. What the racer tries to do is put down a layer of wax whose consistency is just slightly softer than the snow crystals themselves. This way the edges of the crystal penetrate into the layer of wax during the time the racer is pushing off or 'kicking.' Then the bond breaks, and the ski glides forward on a very thin layer of water created by the heat of friction between the ski base and the snow."

Companies are reluctant to divulge the exact ingredients of their waxes, but basically they achieve the different consistencies by blending oil resins of varying grades into paraffin wax. To improve the durability of the wax, they add synthetic rubbers to the formula. Additional substances called microcrystallines allow the wax to bind to the snow. The more microcrystallines added, the harder the wax will be.

More than a dozen waxes are available to match different temperatures of snow. Since weather conditions and snow temperatures are constantly changing, a successful racer must also be a good weatherman. "The coaches and the team spent several days before a major competition evaluating the condition of the snow," says Gaarud. "Then, on the morning of the race itself, they'll spend hours waxing up a variety of different skis for each skier to anticipate the precise condition at the time the gun goes off. Coaches themselves attend clinics sponsored by the major wax companies in which they refine their waxing art. But even though the coaches are relied on for help in waxing skis before a race, the Olympic rules specifically forbid them from helping to wax a racer's skis once the race begins."

Preparing skis for a race is an elaborate chore. First a small blowtorch is used to heat up the old wax so that it can be scraped off with a steel blade. Then a layer of base wax is applied all over the ski to serve as a bond between the polyethylene ski itself and the other waxes. This layer is smoothed with an iron. When the skier is fairly certain of the snow conditions, a glider wax is selected and ironed onto the base wax. The final phase involves putting down a layer of wax directly beneath the foot, the so-called kicker zone. Like a downhill ski, a cross-country ski has a small rise or camber in its middle beneath the point where it is attached to the foot. During the kick, the skier's weight flattens the ski, bringing the sticky "kicker" wax in contact with the snow. When the skier transfers his weight to the other ski to glide, the camber helps transfer some of his weight from the middle section to the other parts of the ski where the glider wax is.

The trick is having the right wax on your ski at the right time. In a long race, for example, the sun might emerge from behind clouds, warming the surface of the snow by as much as twenty degrees Fahrenheit. The wax used to start the race would then become virtually worthless because the ski wouldn't grip the snow at all. "If a racer's wax is wearing off, or if the sun warms the snow, and he's starting to slip," says Gaarud, "then all the coach can do is hand the racer a tube of wax as he goes by. The racer has to step off the course, apply the wax himself, and then start running again." But a racer can anticipate warmer weather by laying down a soft base and covering it with only a thin film of hard wax that will wear off as the race progresses, exposing the softer wax for the finish. Or, if a course begins with a series of hills and ends with a long downhill stretch, a racer might layer his waxes to try to get the kicker wax to wear off at the top of the hills so that his skis will slide more easily when he heads down to the finish.

For those recreational skiers who don't want to spend more time waxing than skiing, there is a whole new market of waxless skis that perform well in almost all types of snow. Most have a fish-scale or wedge-shaped pattern stamped into the polyethylene itself so that the ski grips when the skier pushes off and then glides forward. Some companies have begun experimenting with interchangeable ski bases so that a whole new bottom can be snapped on when the snow conditions change. Most racers, however, will stick to wax. Its infinite variety allows them to literally tune the ski to the snow, making the bottom of the ski as unique as each snow crystal on which it glides.

Building Better Bikes

ERIC PERLMAN

Allan Abbott's entry at the Bonneville Salt Flats "Speed Week" was a very strange bicycle. It had a monstrously oversized, twenty-two-inch diameter chain sprocket—nearly as big as the twenty-four-inch rear tire—no gears, minimal turning capacity, a high-impact bumper, and a hand-built frame designed for stability at high speeds.

A 1955 Chevrolet with a 650-horsepower engine and a tool shed–sized windscreen strapped to the rear roared across the salt flats—towing Abbott on his bicycle. At sixty miles per hour, Abbott announced into his helmet microphone that he had achieved enough speed to begin cranking the chain sprocket himself and was cutting loose from the tow cable.

Car and cyclist accelerated smoothly until the bicycle smacked the rear of the Chevrolet at 110 miles per hour. Abbott wrestled to regain control of the handlebars, called for the pace car to accelerate and streaked through the timed mile. For the last three-quarters of a mile, Abbott averaged 140.5 miles per hour, breaking the world record for the "motor-paced mile" by more than 13 miles per hour.

The world record for bicycle speed—without a pace car to block the wind—is 62.92 miles per hour. This record was set in May 1980 at the International Human-Powered Speed Championships in Long Beach, California. A team of engineers from General Dynamics designed and built the three-wheeled cycle and its aerody-namic fiber glass shell. In a new twist on a bicycle built for two, two champion sprinters mounted back to back—one facing for-ward and one backward—pedaled the vehicle to victory.

This three-wheeled wonder and other vehicles sporting uncon-ventional rider positions are known as recumbents. Though they first appeared at the turn of the century, recumbent bicycles have recently resurfaced in an effort to improve on what is already an amazingly efficient, elegant, and refined design—the standard bicy-cle in use since the early 1900s. The recumbents represent the latest stage in the continuous evolution of the bicycle, which began with those zany shapes of the early days—from the "pedes-trian hobby horse" of 1816 to the "bone-shakers" and "highwheel-ers" of the mid-nineteenth century to the practical geared bike of 1871.

A bicycle and rider is the most efficient moving combination in the world. No organism or powered vehicle requires less energy to move as much mass over the same distance. The superbly hydro-dynamic salmon is the most efficient animal, yet it requires two and a half times the calories to power an ounce per yard as a human on a bicycle. A jet transport plane requires four times the calories; a walking human, five times; a dog, ten times; a helicop-ter, more than two hundred times; and a mouse, six hundred times.

Recumbent designers are challenging the two major obstacles to better human-powered transportation: a somewhat inefficient body position and wind resistance. By lying prone or sitting back comfortably, legs stretched forward like those of a bobsled rider, the recumbent rider rechannels the muscle tension normally used to stay upright into the forward movement of the bike.

Even greater gains in efficiency can be achieved by reducing wind resistance. When moving at a cruising speed of ten miles per hour, cyclists burn up about a third of their energy merely pushing away air. They use more than half their energy to overcome wind resistance at twenty miles per hour; at racing speeds of thirty miles per hour and over, 90 percent of the effort is devoted to neutralizing the effects of the wind.

The response to wind resistance has been the fairing—a light-weight shell that redirects the flow of air around the bicycle. All the recumbents bikes that have surpassed fifty-five miles per hour are encased within fairings that may reduce drag by as much as 50 percent. Unfortunately, the same lines that make a fairing slice neatly through the wind also make it treacherous in crosswinds.

The homemade Bluebell, a recumbent bicycle, was pedaled by Britain's Tim Gartside at 51.9 miles per hour, a world record for two-wheelers. (© John McGrail/Wheeler Pictures)

For this reason the speedy recumbents have been referred to as high-speed accidents looking for a place to happen. To help overcome this instability, some recumbents have added a third wheel.

The reevaluation of bicycling efficiency is not exclusively a recumbent phenomenon. The standard ten-speed's aerodynamic shape has been improved by incorporating handlebar fairings and enclosed front spokes. But standard bikes undergo their most dramatic refinements as they are custom fitted in response to different bicycling events. Modern competition bicycles at $2,000 feature manganese-chromium-molybdenum steel alloys and silver-brazed joints.

Minor adjustments of a bicycle's frame radically alter performance. The front end of a track-racing frame is steeply angled, and the forks that hold the front wheel are curved very little. This combination maximizes responsiveness but transmits a lot of road shock directly to the rider's hands and seat. These bikes are ridden almost exclusively on hardwood or concrete riding tracks in such events as sprint and pursuit racing where very fine control is crucial for cutting in front of the competition at the last moment.

Bikes used in velodrome track events have neither gears nor brakes, and the cyclist's feet are strapped to the pedals. Less hardware lightens the track bike, but the stresses from abrupt turns and breakaway sprints require stronger struts and joints that will not whip or shake. Bicycles for road racing are chosen to fit the course; many hills require wide-range gearing; rough pavement requires a stronger, more shock-absorbent frame and wheels.

In France's quadrennial 750-mile, nonstop Paris–Brest–Paris race, an American team used a new crankset called the Selectocam. It took second, twenty-fifth, and twenty-seventh places out of 1,800 entrants, abruptly ending European disdain for American cyclists and cycle technology. The Selectocam has a standard fifteen-speed gear set, but one of the three front gears has been redesigned to accommodate the quadriceps muscle's natural strengths and weaknesses. At the top and bottom of the power-stroke where leg muscles have the least leverage, the gear offers little resistance, but twenty degrees past the top of the cycle, where legs have the most strength and leverage, the resistance increases due to the shape of the gear. This allows the legs to apply more power where and when it is most naturally available.

The newest design twist for bicycles, however, have been made not for the racer but for the everyday commuter. Mountain bikes—bicycles with upright handlebars and wide tires—have combined the best of the lightweight racing technology with the durability of an off-road bike, making life a little more comfortable for those who would rather pedal than drive. And for those who hate to shift gears, there is even an automatic transmission for bicycles, a fifteen-gear device that shifts in response to pedal pressure.

Indy's Wings of Victory

PATRICK COOKE

Since the Indianapolis 500 premiered in 1911, one of the few elements of the race to remain unchanged has been the objective: The first car to complete five hundred miles (roughly the distance from Indianapolis to Washington, D.C.) takes the checkered flag. The three-hour ordeal is physically punishing, mentally exhausting, and hazardous. Even Ray Harroun, who won the first Indy at a speed of 74.59 miles per hour, accepted his $1,400 prize by telling reporters that he was pleased with the outcome but would never race again—"not for twice the money."

And with good reason. Through the years Indy has been a showcase of tremendous advances in automobile design—but also of accidents, which have come with gruesome regularity. As larger and larger engines (sometimes two under one hood) hurled the racers around the 2.5-mile oval track, lap speeds climbed to nearly two hundred miles per hour. But all that horsepower, essential to winning in the straightaways, made it more and more difficult to keep enough rubber on the road in the turns, and those racers who didn't sometimes found themselves in an airborne posture known as "going over the wall."

In the 1970s Indianapolis 500 officials, concerned with safety, set up an intricate set of engine guidelines, which in effect limited cars to about 640 horsepower. Some racers grumbled, but car designers got to work. Since they couldn't go faster on straightaways

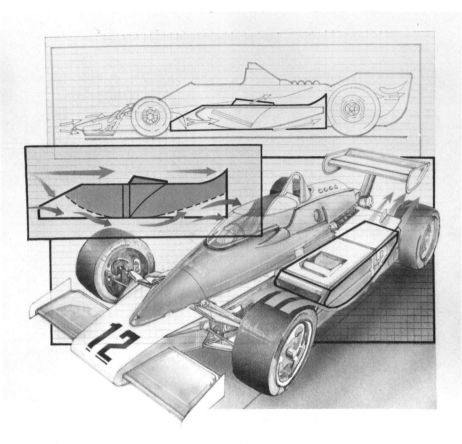

The tunnels on Indy cars are curved on the bottom and flat on top. The air below is forced to travel faster and is thus at a lower pressure, which forces the chassis down onto the track. (© Diversified Arts)

by increasing horsepower, they made it possible for cars to go faster on curves by improving traction. To do this they used air— and a device called the wind wing.

During the 1960s some cars had been outfitted with what was essentially a long, flat, fiber glass plank suspended horizontally from twin goalposts over the rear wheels. When this airfoil's leading edge was angled into the wind, the oncoming air would push the car down, keeping it from skidding around corners.

Air, however, takes on a new personality when something is moving quickly through it. At low speeds it tends to flow over a surface smoothly, like a slowly moving stream flowing around a log. But at high speeds air gets caught on uneven surfaces, forming eddies and, at high speeds, the aerodynamic equivalent of rapids. This causes drag, and it is a formidable problem. The wind wing, so effective on curves, was like a log in a fast stream on straightaways—the added friction and wind pressure caused by the ungainly overhang reduced speeds and wasted fuel.

So designers put the wind to better use. For the last two years all cars racing at the Indy have used the venturi chassis, essentially an upside-down wing mounted on each side of the driver. Except that it has long boxy tunnels stretched along either side of the driver's seat, there is little to distinguish this chassis from those of previous Indy-type race cars. Each tunnel, or pod, forms an inverted wing, flat on top and curved on the bottom. The design takes advantage of the same principle that gives airplanes lift— only in this case the "wing" is flipped to push the car downward. Air moving under the curved surface travels faster in its attempt to keep pace with air rushing over the straight, flat side. This creates a low-pressure area between the ground and the low-slung racer that "flies" it down onto the track. The entire machine becomes a wing, in effect, as it accelerates into the wind. The greater the speed, the more "downforce" occurs. "You can actually feel the car squat down as the tunnels begin to work at relatively low speeds in the seventy to eighty range," says veteran race driver Bill Alsup, who in 1981 achieved the second best record in Indy-type racing.

Still, for all its magic, the new body shell has created some problems for its pilots. As it speeds along at nearly two hundred miles per hour, a race car punches a "hole" through a wall of air, forming a cone of low-pressure air that stretches twenty to thirty yards behind it. Before venturi cars, pursuing drivers had always found this corridor a pleasant place to ride because the lead car not only absorbed the air friction but actually pulled them along through the hole. With the air-dependent venturi design, however, a car in the low-pressure cone begins to gasp for air, losing its downforce. So disconcerting are the effects, in fact, that looting the breeze is used by top drivers as a tactical device. "If you think of air as water," says Alsup, "it's like suddenly being in the wake of a boat." With so many drivers constantly maneuvering for better position, it is a danger that is almost impossible to avoid.

In the early stages of development, engineers found that for the venturi to operate at full strength, air ingested at the front of the pod had to exit under the back of the car. Any leaks out the side curtailed suction. Therefore, a spring-loaded enclosure device called the sliding skirt was hung on each side of the car. The skirt slid along the track's surface, sealing in the low-pressure area underneath. And because the skirt would be in constant high-speed contact with the pavement, it was equipped with a variety of devices from ceramic rubbing tips to industrial diamonds to keep the edge from wearing away. In 1982, however, the United States Automobile Club, the organization sanctioning the 500, ruled that the fringe could not slide across the pavement, reasoning that a car could be thrown out of control if a skirt snagged the track's asphalt surface. In 1983, it went even further, requiring the fringe to be at least one inch above the pavement.

The edict once again put designers under new pressure to recoup the road-holding ability lost when the skirts were raised. Lighter suspension springs and softer tires were tried in hopes that downforce would push the car closer to the road. But the most popular shift in design has come in what Alsup describes as the black art of retunneling, or changing the contour of both wings' curved surfaces to produce more downforce.

So far, gains have been made mostly through trial and error. Each time a new shape is bolted on, however, dozens of compromises may have to be made in other areas of the car, such as the type of tire used, the placement of the transmission, or even how the driver should sit. And the car must remain a legal entry—no longer than fourteen and a half feet long, eighty inches wide, and thirty-two inches high. Lauri Gerrish, an Englishman who has been working on the problem as Alsup's crew chief, says they've been able to save about 60 percent of the downforce but still have a lot to learn in the wind tunnels. "A couple of ideas we've been testing were sent down to NASA for an opinion," Gerrish says, "and they haven't stopped laughing yet." Because no one has come up with just the right modifications, there should be as many variations as there are Indy entries.

Despite the usual chaos any innovation brings to the rule books, venturi chassis race cars have been a blessing to the Alsups, Unsers, and Andrettis who ply their trade at the "Brickyard." In a sport where a $150,000 car can stall due to the failure of a fifty-cent component, they readily welcome any improvement that behaves consistently, especially one that lacks moving parts. After all, even under the best conditions, thundering two hundred laps in heavy race traffic isn't something you could expect to get most people to do. Not for twice the money.

America's Cup:
Sailing's New Tacks

BILL ROBINSON

In the summer of 1983 a sailboat from Australia shook the yachting world by doing something that had never been done before: it won one more race than its American opponent in the final round of sailing's most prestigious race, the America's Cup. What made *Australia II*'s victory so stunning was that its opponent, *Liberty*, went into the race as the latest representative of the longest winning streak in sports history.

Dating from 1851, the America's Cup race is the longest-running international sporting event—longer than the modern Olympics. The race started when a syndicate from the newly formed New York Yacht Club sent the 108-foot schooner *America*, modeled on our fast pilot boats, to England, where she upset a fleet of top local boats in a race around the Isle of Wight and brought home a silver ewer known as the 100 Guineas Cup. It was donated to the New York Yacht Club for international competition, renamed the America's Cup, and foreign yachts were challenged to come over and try to take it away. Roughly every three years since at Newport, Rhode Island, U.S. boats have beaten all comers in the best of seven series. The *Australia II*, however, had something no other boat had used before: a radical new hull.

The twelve-meter sloops raced in the America's Cup are at the windward edge of sailboat design. The Twelves, which have no accommodations, are pure racing machines. Their use is almost entirely confined to closed-course racing—competition around a set of buoys—in contrast to ocean racing to offshore points like Bermuda or Hawaii. The designation twelve-meter comes from the design formula for the class, not the boat's length. The various elements that determine the vessel's buoyancy, sail area, waterline length, and weight can be altered according to the naval architect's ideas, as long as the result is twelve meters.

Though the sixty-plus-foot yachts present curves of elegant simplicity, their shape, like that of all sailboats, is derived only after considering a complex combination of physical forces. The sail, for example, is designed to catch the wind and produce a force perpendicular to the sail. And because the hull is designed so that resistance to sideways motion is greater than resistance to motion fore and aft, the boat moves forward in response to a wind from almost any direction.

When the Twelves first came into use for the America's Cup in 1958, the boats were wooden and had cotton sails. A seaman's eye and the feel of the boat under him, and navigational dead reckoning, governed how the races were sailed. Cotton sails soon gave way to more durable Dacron and nylon, and lighter fittings reduced weight on the mast, making the boats more stable. The big coffee grinder winches used to pull in the sails were redesigned using better gears and bearings.

Other changes in design and materials came gradually and by trial and error. There was experimentation with titanium, boron, and carbon fiber for masts, but these so-called "exotic" materials proved too expensive and were banned. A major rule change allowed aluminum hull construction, with reduced weight for the same strength, in time for the 1974 series.

In 1974 on-board microcomputers also appeared for the first time at Newport. Equipped with sensors that monitor boat speed and wind speed and direction, a microcomputer can tell a navigator if he has achieved the "best speed made good," that is, the maximum speed for a given point of sailing. They can also predict the wind angle for the next leg to aid in the selection of sails. The crew of the *Clipper*, one of the challengers in 1980, relied on a computer affectionately referred to as "Hal" to plot their best course, tell them whether boat speed was fluctuating too much, and provide information for postrace analysis.

All these technological advances may have changed the look of the America's Cup event, but races are still won and lost by the skippers and their ten-man crews. Ted Turner, a true "seat of the pants" sailor, who in the *Courageous* beat out more highly computerized boats in the 1977 series, has said computers are worthless

The Australia II, *above right, drops its spinnaker and slips ahead of the* Liberty, *above left, as it turns for the last leg of the final America's Cup Race. It won the race, and the cup, partly because of its radically new hull design, left. Unlike conventional hulls,* Australia II's *concentrated most of its ballast in a bulb at the bottom and had wings to help make quicker turns.* (© *United Press International, Inc.*)

to him. The skippers use an intricate system of right of way, tacks, and sail selection against a backdrop of wind shifts to outmaneuver an opponent. There are feints and fakes, even to the extent of beginning a damaging tactical move in the hope that another boat will follow. Experienced skippers are adept at blocking and disturbing the wind enough so another boat cannot overtake them. Once in front, the lead boat will try to stay between the wind and the other boat, which will be trying to tack away to find less disturbed air. This leads to "tacking duels" in which the sixty-thousand-pound boats will sometimes tack hundreds of times during the twenty-four-mile course.

In 1983 the New York Yacht Club struck the rule that foreign boats must use technology and materials only from their home country, in effect taking away the technological advantage U. S. boats have enjoyed for years. Two such materials are Kevlar and Mylar—both plastic-type materials that are highly resistant to stretching. A sail made of them can be used through a much wider range of wind strengths without the need for frequent changes, which tend to slow down a boat. Mylar is damaged by folding, however, and must be carefully rolled for storage. Lightweight cobalt steel rods are used for the standing rigging that supports the mast. New materials have also found favor for running rigging, such as mainsheets and spinnaker guys, which are now of Kevlar, a lighter and stronger synthetic rope. Kevlar, nonabsorbent and nonstretching, wears out in a couple of weeks, so replacements are frequent.

Allowed access to these new materials, a record number of foreign boats—single entries from England, Canada, France, and Italy and three from Australia—turned up at Newport to take on the defending skipper Dennis Conner on the *Liberty*. *Australia II* sailed away with a 4–3 victory.

It wasn't new materials that led to *Australia II*'s victory, however, it was her new shape. While the other boats were very close to the classic 12-meter shape, *Australia II* had much of her ballast in a torpedo-shaped bulb at the very bottom of the keel, the first boat in the America's Cup to sport such a design. The Americans were at first skeptical of the new keel. "We've done a lot of work with a bulb in tank tests," said *Courageous* designer Bill Lansan before the race, "and we don't think it's worth the added drag." Now that *Australia II* is the boat to beat, however, it's likely that her new looks will turn up in some of the new challengers, including the American sloop that sails to Perth, Australia, in 1986 to bring the Cup back home.

Hang Gliding: Science in the Clouds

ERIC PERLMAN

When crazed monks and princes leaped from their castles and cathedrals in the Middle Ages gripping undersized wings made of sticks and cloth, they met with little success. Long on faith but short on aerodynamics, the few "tower jumpers" who survived were carted away with no desire to try again.

Today's hang-glider pilots have surpassed the most goggle-eyed dreams of their tower-jumping forebears. From New Hampshire to New Zealand on any day when the winds are right, thousands of pilots assemble their multicolored wings, clip in their flying harnesses, and step to the edges of cliffs, dunes, and mountain peaks. A quick charge into the wind and these modern-day Daedaluses rise into the sky and begin their search for elevator updrafts that can take them miles from Earth.

Like sailboats, hang gliders come in an array of shapes and sizes and perform differently in different winds. In the unforgiving world of the air, function follows form—exactly. The underlying principles of hang-gliding aerodynamics are the same as for any other aircraft. The wing cleaves the air, causing it to pass both over and under the wing's surfaces. Because of the curvature of the upper surface, air passing over the wing must travel farther and move faster than air passing under the wing. This lowers pressure above the wing relative to that below it, creating lift.

The swept-back wings of an F–16 are designed to achieve maxi-

91

mum performance at supersonic speeds. The billowing belly of a kitelike, diamond-shaped Rogallo hang glider performs best at nineteen miles per hour, with a glide ratio of four to one. That is, for every four feet of horizontal flight in stable air, the glider will sink one foot. Modern hang gliders, which cost around $2,000, offer glide ratios up to eleven to one.

The greatest obstacles to flight and lift are gravity and drag. The leading edge of the wing, the pilot, and other surfaces on the glider impede flight by disturbing the air flowing over the glider. The air also forms tiny eddies at the wing's tips and edge, creating more drag and decreasing lift. Because longer wings reduce this turbulence, some high-performance hang gliders now have wing spans greater than thirty feet.

The original version of the modern hang glider was patented in 1951 by an American, Francis M. Rogallo. The National Aeronautics and Space Administration worked extensively with the Rogallo wing in the 1950s and 1960s in its search for a steerable, gliding parachute for manned and unmanned space capsules. Though the Rogallo design never made it into space, word of the wing reached the public.

Early hang-gliding enthusiasts built their wings out of polyethylene and bamboo. These so-called ground skimmers quickly gave way to gliders constructed with Dacron sails and aircraft-grade aluminum frames. With safer, stronger wings, hang-glider pilots of the early 1970s went higher and farther, flying off mountains and soaring on the steady updrafts atop ocean-facing cliffs such as those at La Jolla's Torrey Pines, Oahu's Pali Cliffs, and San Francisco's Fort Funston. This whetted the pilots' hungers still more, and they began to tinker with the basic Rogallo design, lengthening the wings and tightening the sail for better lift. Some scrapped the original shape of the Rogallo altogether and returned to the biplane glider designs of the Wright brothers or the nineteenth-century ribbed wings and tails of Otto Lilienthal, whose last words before dying from a hang-glider crash were, "Sacrifices must be made."

Sacrifices were made in the 1970s as well. As fatalities and injuries increased every year hang gliding became known as "the killer sport." Then in the late 1970s hang-glider manufacturers started using professional pilots to test-fly the gliders first. The industry set manufacturing guidelines, pilots were given flying ratings, and a new safety consciousness arose, with the result that despite an enormous boom in the sport's popularity, deaths from hang gliding appear to be waning.

Meanwhile, a succession of design changes have given hang gliders more speed, maneuverability, and lift. Hang gliders owe much of their expanded range and safety to the advent of the double-surface sail. This innovation, shaped like the one-layered wing it replaced, presents a typical, cambered airfoil to the wind, but at high wind or flying speeds the two layers compress, resulting in less drag and greater stability. Soaring at low air speeds, the sail billows, making a thicker airfoil that provides a higher glide ratio.

Although the international hang-gliding record committee is located in Paris, most official world records are set in the Owens Valley of California on the eastern side of the Sierra Nevada range. Robert Thompson has flown 139.8 miles point to point. New Zealand's Ian Kibblewhite flew 13,694 feet above his launch point. Though not sanctioned in Paris, the world record for duration aloft was set in 1982 when Jim Will spent a record 24 hours and 32 minutes gliding over the Hawaiian island of Oahu.

These world-class glider pilots fly so high by looking low. Micrometeorology, the intricacies of wind circulation, air density, updrafts, downdrafts, and turbulence within the immediate flying region is as crucial to a hang-glider pilot as surf to a surfer. A hang-glider pilot can read the air to a large extent from the landscape beneath it. A dry, grassy field in the sun warms quickly, heats the air, and causes updrafts. A forest absorbs heat, making for downdrafts. A cliff standing in the face of a prevailing wind offers steady, soarable updrafts, but the top or lee side of the cliff is the lair of the deadly "rotor," a powerful, circular downdraft that can slam a glider straight into the ground.

Like water following the contours of a stream bed, air conforms to the surface features of the earth. A gap in a cliff or a cool deep canyon can cause downdraft or turbulence. Mountain peaks can set up a wave pattern in air similar to waves downstream from a submerged rock in water. If a mountain "downstream" from the wind wave is in the right position to harmonize with the wavelength of the wind that has been set by mountains further upwind, a steady updraft forms and holds on the windward side of the peak, offering a soarable wave that can reach higher than 50,000 feet. Since sailplanes have soared to 46,000 feet, hang gliders will surely continue their rise into the stratosphere. The tower jumpers would have been pleased.

Part III

THE BODY

Wrestling with Weight Loss

MICHAEL GOLD

There were lots of tales about wrestlers trying to lose weight, but the story of Charlie Cheek was the one that inspired me. The legendary Cheek would go out and have three Big Macs, two large fries, and several vanilla shakes. He would then retire to the parking lot and, plunging a hand down his throat, adorn the shrubbery with barely digested fast food. Whereupon he would return to the back of the burger line for seconds. Or would it have been firsts again?

As a college freshman struggling to keep off enough mass to wrestle in our team's 118-pound slot, I thought Cheek's strategy both simple and ingenious. After the next match I dined heavily, strolled back to the dorm, and assumed my position before the "porcelain altar," as we clever college kids used to call it.

Nothing happened. Thinking I might not have reached down far enough with my hand, I tried again with a plastic coffee stirrer— but only gagged. A few people who had gathered outside the bathroom began offering suggestions, the most reasonable of which was to drink hot tea with mustard and plenty of salt.

I had four cups. They were vile but not, alas, nauseating. Ironically, the briny brew made me so thirsty that I drank half a dozen bottles of Sprite that night. The next morning I found I had gained eight pounds—much of it Sprite, no doubt, that my salty body had absorbed like a sponge. I looked like the Pillsbury dough wrestler.

97

No lie.

Now a lot of high school and college coaches claim that kind of craziness just doesn't happen anymore. Many of their wrestlers agree, though not as convincingly. "There are a few Neanderthals left," admits Darryl Burley, the 145-pound captain of the 1982 Lehigh University team. "Some wrestlers do take Ex-Lax. Some regurgitate food. Some sweat in the sauna for ungodly amounts of time. But the good guys today don't do the ridiculous things."

Maybe so. But many a wrestler still lives by the traditional, if questionable, belief that by competing in a weight class far below his normal weight, he will be bigger and stronger than his opponent, presumably a lunkhead who has never heard of the strategy. From December through March, wrestlers make themselves ten, fifteen, sometimes twenty-five pound lighter than their normal weights. They may be crude or civilized about it, but in the end they all give up the same thing: a lot of food and water.

"My own biased belief is that there must be some impairment of growth, particularly among the younger wrestlers," says physiologist Charles Tipton of the University of Iowa. "But there just isn't any hard evidence for it." In fact, no one has documented permanent harm of any sort resulting from the ritual of "making weight." Yet many experts condemn it, especially in its more extreme forms. The short-term effects, they say, are scary enough.

Take dehydration. On the day before a match a wrestler loses his last five or seven pounds by sweating them off. He may simply wrestle vigorously in a hot practice room or perhaps run for miles in a rubber suit. He shows up the next day at the weigh-in well shriveled. The volume of blood in his veins is below normal, making his heart beat faster. Less blood also makes it harder for his body to cool off the inner tissues, which explains why his temperature is higher than normal. In addition, though no one has studied the question, Tipton and others suspect that reduced fluid may upset chemical balances within the cells, impairing their ability to make use of energy compounds.

Many coaches, including 1972 Olympic champion Dan Gable of the University of Iowa, counter that wrestlers can replenish their water supplies—and therefore undo any damage—during the locker room feast that begins after the weigh-in. But even if the athletes could force five quarts of water into their shrunken stomachs along with the submarine sandwiches and Hostess Twinkies, their bodies may be unable to redistribute it by match time. According to Tipton, blood tests, urine samples, and tissue biopsies have shown that wrestlers are still quite dehydrated one hour after eating and drinking—and one hour after the weigh-in when high school matches begin. Even the three-hour log before collegiate tournaments is not enough time.

Hard-core dieting creates a number of other complications. Cut off from an outside source of sugar, a wrestler's body begins breaking down its supplies of glycogen, a starch, into glucose. Because every ounce of glycogen is stored with three ounces of water, the process can flush as much as two pounds of water from the body during the first day. And since few wrestlers have any spare fat, the body then begins converting its protein into glucose. A starving wrestler may lose several ounces of protein a day, much of it muscle tissue. According to one study it takes two to three days to fully recover from short-term starvation.

The American College of Sports Medicine finds all this so disturbing that it has concluded that there is no way anyone can medically justify "the weight reduction methods currently followed by many wrestlers." Of course, Charlie Cheek and his disciples never felt the need for a medical justification. Without any hard evidence for permanent damage, what they want to know is whether any of these short-term and apparently reversible effects hinder performance.

On the question of muscular strength, there is conflicting evidence: Some wrestlers who cut weight seem to weaken, some don't. There's no doubt, though, that starving wrestlers are low on energy. Chemical analyses show they have depleted their stores of glycogen, the primary energy source for anaerobic activity, the quick bursts that characterize much of wrestling. Most tests have found that wrestlers become exhausted much sooner when they are fasting. On the mat the phenomenon is known as running out of gas: A wrestler who has been crash dieting appears to doze about halfway through the match and is flattened by a well-fed opponent. Indeed, it isn't unusual for a starving wrestler to fall asleep while awaiting his turn to compete.

On the other hand, Harold Nichols, who has coached at Iowa State for thirty years, found that members of his team who dropped four to sixteen pounds in thirty-six hours actually benefited. The wrestlers developed faster reaction times and better balance; strength and heart rates were unaffected.

In light of the somewhat confusing results of research, the people who advise wrestlers have chosen a middle ground. There is no point in a wrestler's starving himself into a state of exhaustion, they say. If he insists on losing weight it should be done gradually, through dieting and exercise—and much of the loss should come before the season starts. It has been estimated that a daily diet of 1,200 to 1,400 calories could supply the needed protein, vitamins, and minerals and at the same time cause a loss of four to five pounds a week. "No one has yet shown any harm from slowly reducing the caloric intake of a well-balanced diet," says Donald Cooper, head of the student health service at Oklahoma State University.

In fact, in twenty-five years of caring for college wrestlers, Cooper says he's observed only one long-term effect of the quest to make weight. "Many wrestlers put on a few extra pounds when they get out of school." Of course, that might even happen to someone who never spent a weekend in the hotbox or lived for days on nothing but grapefruit.

Winning Recipes in the Athlete's Kitchen

WILLIAM F. ALLMAN

Popeye was almost right. When he wanted to be "strong to the finish," the mighty mariner reached for his spinach.

He would have been better off with spaghetti. Spinach contains carbohydrates, the stuff that makes athletes go. But pasta has a lot more. That's why many a runner stuffs down as much of the floury food as possible the night before a marathon.

Loading up on carbohydrates is the latest dietary approach to getting extra energy for that extra mile. The old standby, a training meal of steak and potatoes, still graces the dining tables of many college football teams, but some professional athletes prefer more exotic concoctions. Washington Redskin lineman Dave Butz drinks two milkshakes before a game; BoSox slugger Wade Boggs eats chicken.

Athletes may put almost anything into their mouths before a game, but the main fuel that moves their muscles during it is a chain of carbon and hydrogen called glycogen. And carbohydrates are the main source of glycogen. "If you think of your body as an engine," says William Fink, a sports physiologist at Ball State University's Human Performance Laboratory, "then carbohydrates are the gas." Be it pasta, potatoes, or pancakes, their carbohydrates are stuck together as glycogen and stored in the muscles.

During strenuous exercise glycogen is broken down, releasing energy that is used to make muscles contract. Fat can also fuel muscles, but it requires more oxygen to break down. The exact ratio of fat to glycogen burning depends on the individual and the intensity of the exercise, but fat alone cannot supply enough energy for an athlete. "There's an old saying the fat is burned in the fire of carbohydrates," says Fink. "It's not like you can switch from one tank to another."

Protein, too, can be burned for fuel, but it is even less efficient than fat. The main reason athletes need protein is to build muscle tissue. Therefore, many weight lifters and football players feel they need to eat a lot of meat or take protein supplements. But according to Nancy Clark, a nutritionist at Sports Medicine Resource, Inc., in Boston, the American diet is already so full of protein an average person eats more than enough to satisfy the needs of an Olympic weight lifter. And while red meats contain a lot of protein, they also contain a lot of fat and cholesterol, neither of which helps an athlete. Protein supplements, like many other additives thought to give an athlete an edge, simply leave the body as urine.

According to Clark, athletes need about 4 grams of protein for every 10 pounds of weight. Therefore, a 150-pound athlete should take in about 60 grams of protein a day—an amount easily supplied by a piece of chicken, two eggs, and a cup of milk. Still, says Fink, "If the coach takes a football player out for a meal and the player is given a choice between macaroni and cheese and filet mignon, most will take the steak. But the macaroni is better for them."

Carbohydrates are made by plants, which use photosynthesis to combine carbon, hydrogen, and oxygen into sugars such as those found in sugar cane and fruit. Beans, grains, potatoes, and vegetables contain longer chains of these sugars called starch. But because the carbohydrates in leafy vegetables such as lettuce are mostly in the form of cellulose, which is not digestible, Popeye would have to eat quite a few cans of spinach to get all the fuel he needed for a long bout with Bluto.

Running out of fuel is a problem for athletes such as marathoners who exert themselves for more than two hours. Glycogen stores are limited, and when they run out in the final miles of a marathon, a runner experiences what is known as "hitting the wall," a sudden onset of tremendous fatigue.

One way to push the wall back is carbohydrate loading. Many marathoners, cross-country skiers, and bicyclists reduce the quantity of the carbohydrates they eat several days before an event while continuing to train. This depletes the stores of glycogen. Then, the evening before the race, the athlete loads up on pasta, bread, and other carbohydrates. The glycogen-starved muscles rebound, soaking up more glycogen than usual.

Some physiologists doubt it's necessary to deplete the glycogen stores first. An athlete's muscles can store twice as much glycogen as those of a sedentary person. Some athletes, like marathoner Bill Rodgers, simply eat more carbohydrates a few days before the event.

Loading up on carbohydrates only works for athletes involved in very strenuous exercise for long periods of time—not for someone getting ready for the company softball game. "Carbohydrate loading is only effective after two hours of activity," says Fink. "Many people just use it as an excuse to eat a lot of spaghetti, garlic bread, and beer."

Vitamins are another food that provides the mirage of a competitive edge. Vitamins are co-enzymes that help carry out many of the body's chemical reactions such as calcium absorption and protein synthesis. The body does not manufacture any of the thirteen known vitamins except trace amounts of vitamin D, so they must come from food or supplements. But "there is a feeling among some athletes that if a little bit of something is good, then eighteen thousand times that amount must be better," says Jackie Puhl, sports physiologist for the U.S. Olympic Committee. "It is possible that athletes need more vitamins, but the large amounts of food they take in usually takes care of it."

Another athlete's food myth is that candy bars supply "quick energy." According to Fink, though candy bars release sugar into the bloodstream, there is already enough there for several hours of exercise. A candy bar, like most foods, does no good until it has been processed by the body—the day after it's eaten. In fact, says Fink, eating candy right before competition could cause trouble. The rise in blood sugar stimulates the release of insulin, which in turn removes glucose from the blood, resulting in an overall *decrease* in the blood sugar level that can make an athlete listless and tired.

It might be a better idea to reach for a cup of coffee instead. David Costill of the Human Performance Laboratory found that after drinking two cups of coffee athletes could do strenuous work for fifteen minutes longer. In another study a group given the caffeine in about a cup and a half of coffee pedaled a bicycle an average of 7 percent harder but perceived the effort as being no different than usual. It's not caffeine's power as a stimulant that boosts endurance; it's that the caffeine helps release fats into the blood, causing the body to burn glycogen more slowly.

Of course, too much caffeine may make an athlete jittery and distracted, just as too much pasta may cause indigestion. Nutritionists recommend that athletes do what feels best. There is no overall nutrition program for Olympians, says Puhl, and athletes have to eat so much to meet their caloric needs anyway that an exotic foray into unconventional foods generally does not hurt. Marty Liquori, for example, reportedly had a stack of pancakes and a glass of wine just before winning the gold medal in the 1,500-meter race at the Pan American games in 1971. "Basically, a balanced diet is the most important part of nutrition," says Anne Grandjean, nutritionist and consultant for the Chicago White Sox and the Olympic Committee. "We don't see many strange diets among Olympic athletes. If a person had a really weird diet, he wouldn't be there."

The Physiology of Keeping Cool

SUSAN WINTSCH

Humans, like automobiles, are water cooled. During the 26-mile 385-yard marathon a runner may sweat as much as a gallon of water, more than 5 percent of total body weight, in an effort to regulate the temperature of the human engine. Sometimes the efforts are in vain: The number of victims stricken with exhaustion and heatstroke rises with the thermometer.

To beat the heat and dehydration athletes grab cups of liquids on the run or during a break. Just what is in those cups has grown into a multimillion-dollar enterprise and stirred a scientific debate that has generated some heat of its own.

In 1968 the Stokely Van Camp, Inc. food company introduced Gatorade, a lemon-lime drink with a lemon-lime color. The company claimed that Gatorade's unique "isotonic" formula, having the same solid-to-liquid ratio as body fluids, would replenish fluids and salts, provide extra energy, and enter the system "twelve times faster than water."

Developed for the University of Florida "Gators" by university physiologists headed by Robert Cade, Gatorade received the endorsement of the NFL, NBA, and major league baseball and soon became the drink of choice for sports teams across the nation. "If it works on the thirst they work up," ads ran, "think what it can do for you and the kids."

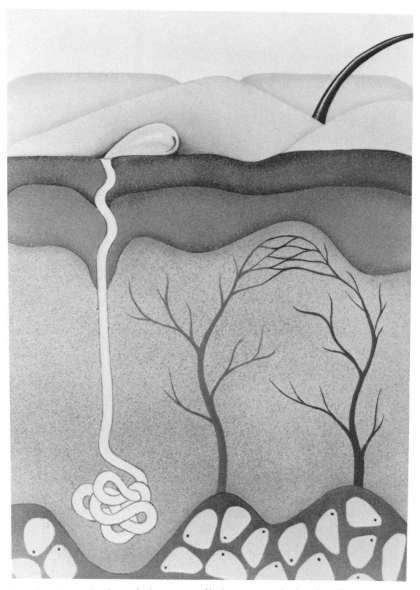

Drawn from the liquids between cells by a special gland, a drop of sweat cools the body as it evaporates off the skin. (© M. E. Challinor/Science 82)

A host of isotonic competitors followed, including Bulldog Punch, Thirst Quencher, Sportade, and QuicKick. Athletes, presented with a bewildering array of alternatives to the proverbial water bucket, weren't quite sure what they should be drinking.

Because body fluids are depleted through sweating more rapidly than they can be restored through drinking, the competing demands for body water during activities like a marathon can be a strain. Water is drawn from the blood plasma into the muscle cells to dilute the by-products of metabolism. At the same time, an athlete uses water in the form of sweat to dissipate heat. The release of sweat draws fluids back from the cells and spaces between them into the plasma. The hotter the body is, the more blood flow to the skin competes with supplies to the working muscles. Unless the athlete drinks large quantities of fluids, these physiological tensions can become critical, halting perspiration and culminating in heatstroke.

Gatorade's initial claim of an absorption speed twelve times faster than water was "pure nonsense," according to physician Gabe Mirkin. The claim, for which Mirkin says there is "scanty evidence," was first challenged by a 1970 study at Ohio State University that showed no significant difference between the absorption rate of Gatorade and of water taken by subjects previously dehydrated in a sweatbox.

In fact, Gatorade might even be slower to reach the bloodstream. Since the intestine absorbs liquids rapidly, the key to replacement is how soon fluids leave the stomach, where very little absorption takes place. David Costill, director of Ball State's Human Performance Laboratory in Indiana, has found that the single ingredient having the greatest effect on the rate of stomach emptying is sugar. He found that Gatorade, which has a high sugar content, leaves the stomach less rapidly than water. Athletes who prefer the taste of Gatorade to water or who drink it specifically for the energy provided by its carbohydrates can make it easier to absorb by mixing it with water.

While the dangers of excess water loss during exercise are understood, the implications of the accompanying loss of minerals and salts, or "electrolytes" are still uncertain. Electrolytes are normally dissolved in the body's fluids, where they regulate the distribution of cellular and extracellular water and help control pH. Rich in sodium and chloride, sweat also contains lesser amounts of potassium, magnesium, and calcium. Isotonic drinks, according to their manufacturers, are beneficial because they quickly put back what the body loses in sweat—its minerals and salts as well as its fluids.

Many physiologists, however, are not convinced that this electrolyte bonus is necessary. Costill's studies indicate that an individual who sweats away nine pounds of fluid will only lose roughly 6 to 8 percent of the sodium and chloride and only about 1 percent of the potassium in body fluids. And he found no evidence that drinking electrolytes during exercise improves performance. Electrolyte beverages can be useful in replacing minerals and salts after exercise, but, says Costill, the same thing can be accomplished by eating fruits and vegetables.

James Wilkerson, a physiologist at Indiana University and a former colleague of Costill's at the Human Performance Lab, cautions against taking lots of extra electrolytes during exercise, particularly since the release of potassium into the blood (which widens the blood vessels to aid heat dissipation) and the production of sweat (which removes more water than electrolytes from the plasma) makes the concentration of electrolytes in the blood progressively greater. The sudden influx of the extra sodium, chloride, and potassium in a salt tablet, Wilkerson warns, can confuse the kidney, causing it to "overreact" and excrete more salts and minerals than are ingested. Physicians, well aware of this problem, emphatically warn athletes against using salt tablets as they exercise.

The body's ability to compensate for electrolyte losses during successive days of dehydration also argues against a special need for replacement beverages. After repeated periods of heavy sweating, the blood begins storing extra sodium and water, and the kidneys excrete fewer salts and minerals into the urine. In fact, Costill's attempt to induce a potassium deficiency in subjects by feeding them an "absurd" low-potassium diet over several days of heavy exercise was unsuccessful.

But James Knochel, a physiologist at the University of Texas's Southwestern Medical School in Dallas, examined six military recruits during a five-week regime of basic training in hot weather and found that half the men became potassium deficient. Initial symptoms of this condition include unusual weakness and listlessness. In cases of severe deficiency, muscular performance may be impaired and tissues may deteriorate. Other scientists have also reported lowered blood potassium in athletes after a season of training, indicating a need in some individuals for supplements beyond normal dietary intake. Beverages with additional electrolytes, if not taken during exercise, are fine for replacement, says Mirkin, but so are a variety of fruit juices.

Sports medicine experts have thus tended to downplay electrolyte replacement in their advice to athletes. "Concentrate on what is essential—water," George Sheehan, sports medicine author, tells his readers. Sheehan maintains that even beer, a diuretic, can satisfy the runner's immediate need for fluid and energy.

Since thirst is generally quenched before very much fluid is consumed, athletes are usually encouraged to drink as much as they can of whatever tastes good and goes down easily. Apparently that advice is followed by people mowing lawns, weekend sports participants, and armchair athletes as well—they've helped create a $75 million market for Gatorade and other isotonics. Many NFL teams still use the drink. And yes—so do the Florida Gators.

Weight Lifting:
What Makes Muscles Work

WILLIAM F. ALLMAN

Long before the invention of barbells, Nautilus machines, and Cybexes there was Milo of Croton. A champion wrestler in sixth-century B.C. Greece, Milo hit upon a great idea for training. Legend has it he got himself a newborn calf and carried the beast around on his shoulders every day. By the time the calf had grown into a bull, so had Milo. He won the Olympic wrestling title six times.

In Milo's day strength was handy in everyday life. Since the invention of the fork lift, however, yanking a hunk of metal away from the earth and standing underneath it for a few seconds has lost some utility. So it has become a sport.

Despite its apparent simplicity, or perhaps because of it, people have different ideas about what the sport of weight lifting should be. Some confine themselves to seeing how many pounds they can lift. Others feel that strength should be combined with speed. Still others think that what's most important is not what they lift but how their muscles look after they lift it.

Regardless of which form the lifting takes, the main ingredient in training for it is muscle building. The basic tenet of any training regime varies little from that of Milo's: The only way to build muscles is to make them work harder than they're used to. And the only way to make them work harder is to make them move more and more weight.

110

But a muscle isn't as simple as the piston on a fork lift. Skeletal muscle, which is the prime mover in weight-lifting events and comprises some 40 percent of a man's body weight, looks different under a microscope than the muscles that wrap the organs and form the heart. It's striated, striped with tiny dark bands.

A muscle is made of thousands of fibers, some as much as a foot long, that run alongside each other like wires in a telephone cable. By and large, the number of fibers remain constant from birth— the 97-pound weakling has approximately the same amount he would have if he were a beach brute. Inside each fiber are short cords of protein linked end to end like boxcars. Anchored at each end of the boxcar and extending toward its middle are thin protein filaments called actin; in the middle of the boxcars are thicker filaments called myosin. The actin filaments don't reach all the way to the center of the boxcar and the myosin doesn't reach to the edges. Where the two fibers overlap they form a dark band. Because the filaments are all oriented the same way, the dark bands coalesce into dark lines across the muscle.

Exactly how these two types of protein conspire to lift a barbell is still obscure, but the prevailing theory says that as a weight lifter tugs on a weight the muscle cells release calcium. The calcium in turn activates cross bridges that stretch between the myosin and actin filaments. The bridges "pull" like rowers in water, causing the myosin fiber to slide past the actin fiber. The cross bridges then reattach themselves and pull again. There are some two hundred oars for every myosin boat and the action is repeated many times a second. "But they don't all work at once," says Roger Cooke, a biochemist at the University of California at San Francisco. "If they did, the muscle would jerk." Instead, only some of the cross bridges pull at any one time. The energy involved in muscle contraction works to break these cross bridges, not connect them, which is why the muscles stiffen in rigor mortis after death.

Actin and myosin are continually being broken down by enzymes—something that is familiar to anyone who has had a leg in a cast for a while. "The half-life of muscle tissue is short," says Cooke. "All the muscle tissue gets completely turned over within two weeks." Of course, muscle tissue is also constantly being built up, and it is this process that weight lifters try to accelerate.

This happens naturally during adolescence in males, thanks to the hormone testosterone, which appears in young boys to hasten the muscleworks. Some athletes think testosterone and its altered forms known as anabolic steroids might still work on the adult body and so take huge amounts of the drug. But the evidence of their effectiveness is mostly anecdotal, and they are also suspected of producing harmful side effects such as liver and testicle damage. Steroids are banned in all international competition and are stringently tested for in the Olympics and other meets.

The more conventional route to a gold medal is pumping iron. When a weight lifter repeatedly stresses a muscle, it responds by growing more actin and myosin. The fibers get thicker, more blood vessels grow to service them, and the muscle gets bigger and stronger.

In the old days when you went down to the local gym to pump iron you got a concrete floor, a strong scent of sweat, and a couple of friends to make sure the dumbbell you were trying to hoist didn't break your neck. Now gyms are a dying breed. In their place are spas, padded and carpeted, and though in some you may see a few free weights relegated to a corner, the most commanding feature in the room are huge machines that look a little like Rube Goldberg jungle gyms.

These are variable resistance machines such as the Nautilus and the Universal gym. They are an attempt to correct some of the problems of free weights. For example, lifting a free weight over the head exercises a lot more than the arms—it works the back and thighs and requires balance. Free weights also allow weight lifters to "cheat" on some exercises: bending the back, for instance, to help with an arm exercise. The biggest problem, however, is that muscles are not uniformly strong. That is, the joints that a muscle moves are capable of more leverage in some positions than in others. This means that a lifter must hoist only the amount of free weights he can safely move through the part of the exercise where his muscles are weakest—the first ten inches of a bench press, for example. During the rest of the lift, however, his muscles are not working as hard as they can.

Variable resistance machines hold and guide the weights, making lifting safer and allowing a weight lifter to isolate a specific muscle or strengthen a particular movement of several muscles. But their main advantage, say their advocates, is that they are designed to push back harder at those points in an exercise where a muscle is capable of pulling harder. The Nautilus machines achieve this by linking the weights through a "gear" that gets progressively wider. It looks a little like a nautilus shell. Other machines, such as the Cybex, use fluids to vary resistance.

While the people who make the Nautilus claim twenty-three out of twenty-eight National Football League teams as devotees, most top-notch weight lifters—including the United States Olympic team—are sticking with free weights. "The problem with the machines is that while you can vary the weight, you can't vary the motion. And a setting that might help a basketball player isn't going to do much for a runner," says Michael Stone, an exercise physiologist at Auburn University and consultant to the Olympic team. "I haven't seen one research article that supports the Nautilus training theories."

John Garhammer, an expert in biomechanics who runs a lab at St. Mary's Hospital in Long Beach, California, agrees: "To the average person on the street the machines are fine. But it depends on what you're trying to accomplish. It's better to train so multiple muscle groups work together for sports where you need balance."

Weight lifting is one such sport. Power lifting, a variation on weight lifting, requires great strength and has fairly simple motions: In one event, for example, the lifter starts with the weight on his shoulders and does a squat. A weight lifter, on the other hand, must lift a weight from the floor to over his head in one motion, called the snatch, or lift the weight to his chest and then overhead, called the clean and jerk. "It's a very beautiful thing to watch," says Harvey Newton, coach of the U.S. Olympic Weight Lifting Team. "It takes balance, coordination, flexibility, as well as muscles. Body builders probably couldn't do it even if they had the strength."

According to Newton, "power" lifting is a misnomer. Power is the application of a force over time. A power lifter may take several seconds to make a lift, he says, but a weight lifter takes a quarter ton of weight from floor to overhead in about three-quarters of a second. Garhammer has studied films of Olympic weight lifters and finds that during the lifting phase some athletes produce more than six horsepower—the highest power output of any human activity. After that, of course, the rest is easy. "Once you get your arms locked," says Stone, "you can hold up the world."

Architecture of the Knee

WILLIAM F. ALLMAN

If God had intended for us to become athletes, he would not have given us knees. These infamous joints, a tenuous marriage of bone, cartilage, and sinew, are perfect for a stroll in the park. But they are remarkably ill suited to a brief encounter with the likes of a Dick Butkus or a Jack Lambert.

Every year surgeons remove some 52,000 pieces of cartilage from knees that were previously propelling someone down the gridiron. Knee injuries make up about 75 percent of serious injuries in football. "Football isn't just a contact sport, it's a collision sport," says Edward Percy, an orthopedic surgeon and head of the University of Arizona's Sports Medicine program. Even so, less than half the knee injuries in football are caused by a direct blow to the knee. Most arise during the tangling of bodies that results from tackling or blocking. Twenty percent come about with no contact at all.

Probably the most infamous knees in football are those of former Jet quarterback Joe Namath. During his thirteen-year pro football career, Broadway Joe had two ligaments and nearly all the cartilage removed from both knees in four operations. And Joe was just a quarterback. It's worse for running backs: Their chances of having knee surgery are about 50–50.

The knee's vulnerability to sports injuries is not the result of some intrinsic flaw in its design. In fact, the knee is a nice compromise between mobility and stability. The shoulder, for example, is highly mobile, making it ideal for pitching a baseball, but such mobility also makes it easy to pop the shoulder out of its socket. The hip, on the other hand, is not as flexible but it's highly stable: Virtually the only way to dislocate it is to break the bone. The knee is somewhere in between. It can bend 150 degrees, swing from side to side, and twist on itself. And it can absorb the force, equal to nearly seven times an athlete's weight, that occurs as a tight end makes the final cut of a down-and-out pass pattern.

The knee is the meeting place of the two major leg bones: the thighbone, called the femur, and the shinbone, called the tibia. The bottom end of the thighbone is shaped like a baby's behind and coated with tough, rubbery cartilage. The top of the shinbone has two shallow scoops hollowed out of it and a ridge along the middle. The femur rides along this ridge like a cowboy in a saddle. Wedges of cartilage line the top of the tibia, acting as shock absorbers and keeping the femur from rocking too much from side to side. "If you put a billiard ball on a hard surface," says orthopedic surgeon Bruce Reider of the University of Chicago, "all the ball's weight goes through a tiny area of the table. But if you cup the ball, the weight is distributed more evenly. That's what the cartilage in the knee does."

Cartilage can be torn if it's twisted too tight. For example, if a tennis player lunges to return a volley, she may bend her knee to the inside, as if she were knock-kneed. This puts pressure on the cartilage between the outer parts of the femur and the tibia. As the leg twists, the femur pushes even harder. The action is not unlike that of a mortar and pestle: The cartilage is ground between the two bones.

Because cartilage has virtually no blood supply of its own it often cannot get enough nourishment to repair itself after damage. Once it's torn, it stays torn. If the torn cartilage is left in, the knee may function normally, but it may rip more during subsequent activity. A piece may break off and get stuck somewhere else in the knee or grind against the femur or tibia. Most doctors recommend removing all the cartilage or at least the damaged part.

Five years ago such an operation meant a stay in the hospital, five weeks on crutches, weeks of therapy, and a lovely scar along the inside or outside of the knee. But a new instrument now shortens both the recovery time and the scar. Called an arthroscope, it enables some patients to leave the hospital in as little as a day, spend only a few days on crutches, and begin running after two weeks of therapy.

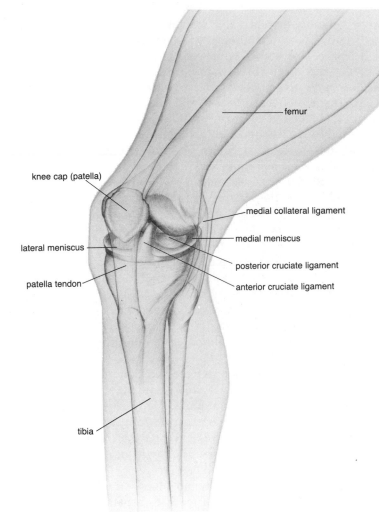

femur

knee cap (patella)

medial collateral ligament

medial meniscus

lateral meniscus

posterior cruciate ligament

patella tendon

anterior cruciate ligament

tibia

Nearly half the running backs in the National Football League will suffer knee injuries during their careers, many involving the straps of tissue, called ligaments, that hold the knee together. Three ligaments are shown in the diagram; the other four are at the back and side of the knee. Of the ligaments, the anterior cruciate and side ligaments are damaged most often. Damage to the menisci, the shock-absorbing cups of cartilage at the top of the tibia, occurs in some fifty thousand football players a year. The injury figures are not likely to be different in the near future. "If you changed the rules of football so that knee injuries were eliminated," says James Garrick, an orthopedist who monitors knee injuries in the NFL, "you wouldn't recognize the game." (© M. E. Challinor/Science 83)

Essentially a tiny lens and light mounted on a pencil-thin shaft (and usually hooked up to a television camera), the arthroscope allows a surgeon to peer into the knee through a small slit in the skin. Cartilage does not show up on X rays, but a doctor can examine it with an arthroscope. The surgeon then inserts equally streamlined knives, scissors, and other instruments to remove cartilage—leaving only tiny scars in the process.

Many athletes can compete without some or any of their cartilage, though the knee is generally more unstable. Without the cartilage's shock-absorbing help the knee will tend to be sore after a game, and the increased wear and tear may lead to arthritis later in life.

An athlete's career is in much greater jeopardy if he tears a ligament. The knee is but a balancing act without the tough, fibrous ligaments that grip the bone like leather straps on a hinge and hold it steady. "The knee has minimal bony stability," says Reider. "If you cut away the ligaments, the joint falls apart."

There are five ligaments around the knee and two more buried deep within the joint. Most knee sprains involve the ligaments at the side of the knee because collisions usually occur at the outside of the leg. If the blow is hard enough, the ligament will tear.

Ligaments are like taffy; once stretched, they stay stretched. "If the knee were meat, the ligaments would be gristle," says Reider. "It's very tough but only stretches 6 percent of its length before it breaks." A partially torn ligament will generally heal itself, but a completely torn ligament must be sewn back together. If the break is in the middle, that can be a difficult task. "Ligaments consist of many strands," says Percy. "It's like trying to sew two hairbrushes together."

The sooner the severed ligament is repaired the better. The fluid in the knee will break down tissue, and if the injury is not repaired within ten days, says Tab Blackburn, a physical therapist and trainer at the Rehabilitation Services of Columbus, Inc., in Georgia, "The fluids in the knee will turn the ligament to mush."

The frontmost ligament, called the anterior cruciate, runs from the front of the shinbone to the back of the thighbone. Doctors are not quite sure why, but this ligament is stretched or torn in nearly 70 percent of all serious knee injuries. Unfortunately, this ligament also has a poor blood supply and it usually does not heal even if sewn back together.

Some athletes continue to play without this ligament. It depends on the job the knee has to do. "A football player who has built up the muscles and other ligaments around his knee might not have any difficulty coming back," says Blackburn. "But a wiry gymnast with a lot of flexibility in her knees may have trouble." If a knee remains unstable, one solution is to jury-rig it with new "ligaments" fashioned out of pieces of other tissue. A piece of tendon from the thigh, for example, is sometimes cut out and attached with staples to the bone around the knee. Percy and his colleagues are experimenting with weaving collagen, a fibrous substance, into a band and using it to bind the bones together.

Knee specialists blame overtraining for the majority of knee problems afflicting amateur athletes, particularly those who are just beginning an exercise program. An estimated 40 percent of all women runners, for example, have knee problems severe enough to require a doctor within the first three months of training. One problem—actually a hodgepodge of muscle, tendon, and bone irritation sometimes called runner's knee—affects nearly a third of the estimated fifteen million joggers in the United States. During running, each foot hits the ground 750 times a mile. Unlike walking, where both feet are on the ground 30 percent of the time, a runner's knee must constantly bear the full load of the body hitting the pavement, a force that equals about five times a jogger's weight. This sometimes leads to a chronic pain behind the kneecap. "Nobody really knows where the pain comes from," says David Johnson, an orthopedic surgeon and consultant to the President's Council on Physical Fitness and Sports.

Orthopedists are hard pressed to suggest a way they would improve on the design of the knee if they could. "We can repair 80 to 90 percent of damaged knees so they're functional again," says Blackburn. "But they'll never be as good as the good Lord made them." Times are tough for the knee because interest in athletics —particularly football and running—is on the rise. The knee's position in the body doesn't help much, either. As the main joint between an athlete and the pavement, it takes most of the punishment brought on by athletics. "The knee is vulnerable because it carries the body's weight," says Reider. "If we all ran around on our hands, we'd be seeing a lot of elbow injuries."

Growing Pains
for the Young Athlete

ELIZABETH STARK

In 1978 Bucky Cox of Lawrence, Kansas, completed a marathon in 5:29:09. Although it was not a very impressive time, Bucky broke a national record—for marathon runners under the age of six.

An estimated twenty million children and adolescents in the United States are participating in some sort of organized sport each year. The average age of many world-class athletes—especially gymnasts—has become so low that many are considered past their prime by sixteen or seventeen. The current fitness mania helps foster the surge of competitive athletics among children, but doctors are worried about the long-term wear and tear on kids' bodies. Thirty-one percent of all sports injuries occur in children between the ages of five and fourteen.

"Kids are not little adults," says Bernard Cahill, an orthopedic surgeon and director of the Great Plains Sports Medicine Foundation. "Their skeletons cannot withstand the same sort of stress."

Children are vulnerable because their bones are still growing—at the ends of the long bones, the joint surfaces, and the places on the bones where tendons attach. These growth areas are made of cartilage, which is two to five times weaker than regular bone and does not harden completely until a child is well into the teens. A fracture in a growth area can be serious: One limb can grow longer than another and bones can grow crooked or stop growing altogether.

119

"A generation ago kids played the sport of the season," says Arthur Pappas, head of the orthopedics department at the University of Massachusetts. "Now they are becoming active in one or two sports where they just do the same thing all year."

Lyle Micheli, director of the division of sports medicine at Children's Hospital Medical Center in Boston, agrees. "At camp kids used to play lots of sports—archery, swimming, canoeing, volleyball. Now children go to camps that specialize in one sport and spend six to eight hours a day at the same activity." The repeated swing of a young tennis player, the continual pounding on the legs of a ten-year-old marathon runner, the constant arching of an eight-year-old gymnast's back all can lead to what are called stress fractures.

Originally called march fractures because they occurred in the feet of soldiers who trained too hard, stress fractures are tiny cracks in the bones and joints caused by the strain of a repeated motion. These tiny fractures often don't show up in X rays and can be misdiagnosed as sprains, tendinitis, or shin splints. And if a doctor prescribes anti-inflammatory drugs that mask the pain, the athlete may return to the sport before the injury has healed. "Sometimes when there is significant damage," says Pappas, "a child is done with a sport forever."

The classic example of a stress fracture is "Little League elbow." The repeated pitching motion, especially when a minor is trying to throw a major league curve, places tremendous strain on the bones of the elbow joint. Little League pitchers are now restricted to pitching only six innings a week and cannot pitch two days in a row. Other common stress fractures include swimmer's shoulder, runner's knee, and spondylolysis, a fracture often seen among gymnasts of one or more vertebrae at the base of the spine.

Hyo Sub Yoon at Rensselaer Polytechnic Institute is working on an acoustical technique to diagnose the elusive stress fractures. When a fractured bone is stressed, it emits sound waves—inaudible to the human ear—different from those of healthy limbs. Yoon stimulates stress in a patient's injured bone by sending ultrasonic waves through it and then records the sounds the bone emits. By comparing these sounds to those of the patient's healthy limb, Yoon can locate the microfractures. If Yoon's technique checks out, doctors will be able to determine if and when an athlete with a stress fracture is ready to return to a sport.

Kids are especially susceptible to injuries between ten and sixteen, the years when the cartilage is growing the most and is weakest. Pappas suggests that children avoid collision sports like football or hockey during this time. Another danger they face then is inflexibility. Because tendons and ligaments usually lag behind skeletal growth, a stretched tendon spanning a joint can pull off a piece of the developing bone if jerked or twisted. These injuries, called avulsion fractures, also occur when developed muscles are stronger than the growth cartilage they are attached to. The pelvic bone is the most frequent site of avulsion fractures since the muscles around the hip are quite powerful and can easily pull off a piece of cartilage during a sharp kick, jump, or turn. Young athletes participating in sports such as figure skating and soccer that involve abrupt twists and jumps are most prone to avulsion fractures.

One of the most common childhood injuries, Osgood-Schlatter's disease, is similar to an avulsion fracture. It often occurs among athletes with very strong thigh muscles. Tiny fractures result from repeated stress on the tendon that attaches the front thigh muscle to the bump of developing cartilage two to three inches below the kneecap. The stress is not strong enough for the tendon to pull off a piece of the bone, but it can be quite painful.

Pappas warns that just because a child looks big he or she shouldn't be thrown into a rough contact sport. "Some kids are larger," he says, "but are skeletally still immature." If children do play collision sports they should be grouped by skeletal age, not by chronological age or by weight alone. "A high school football player with a lot of bulk, no muscles, and loose joints is a sitting duck for a catastrophic knee injury," says Cahill. Some doctors observe the sexual changes that occur during adolescence to estimate a child's skeletal maturity. A more accurate method is to take X rays of the wrist and hand, which have twenty-nine individual growth centers. By analyzing the rate of development at these growth centers, doctors can determine relative skeletal age.

For now, parents and coaches are left with few hard-and-fast rules about what sports are good for what kids. A general belief is that loose-jointed children should avoid contact sports and that children with tight joints should not participate in activities such as ballet and gymnastics. A doctor can determine flexibility and strength during a medical exam.

No matter what kind of joints children have, most doctors believe that they should not run in marathons until their mid-teens. "In the long run, marathon training at an early age ends up being more detrimental," says Bill Dellinger, the distance running coach for the 1984 Olympics. "The longer you wait the better off you're going to be. Running should be fun for kids. Teach them running techniques by letting them play games like soccer that use those skills."

The most important thing, Pappas says, is to let kids enjoy sports while they're young and not force them into highly competitive training too early. Unfortunately, the clichéd image of parents pushing their children into sports against their will remains all too familiar. "Parents should listen to their kids, especially if they complain of pain," says Pappas. "Kids who are performing more out of fear don't do as well."

World Records:
The Limits of
Human Performance

SUE HOOVER EPSTEIN

"World Records are only borrowed," middle distance runner Sebastian Coe observed in 1981, shortly after setting a new mile mark of 3:47.33.

Coe ought to know. The twenty-seven-year-old Briton has checked out eight world records since 1979, three of which still stand (the 800 meters, the 1,000 meters, and the mile). Chances are he'll borrow a few more before he hangs up his racing flats. Ideally, he says, he'd like to run a mile in under 3:46, which is 1.3 seconds faster than his current record.

Research physiologist Jack Daniels thinks it's possible, and with Daniels that's more than just a hunch. With his trusty computer, he looks at athletes' times at different distances, estimates how their speed deteriorates over distance, and considers their strength, training, and physical characteristics such as stride and ability to use oxygen. Then he predicts their future performances. Because of Coe's sizzling 800-meter record (1:41.73), for example, Daniels predicts that the men's mile record will soon be lowered to below 3:46.

Daniels' predictions have a track record of their own. In 1981 he predicted that Alberto Salazar would break the world marathon record, which had remained inviolate at 2:08:33.6 for twelve years. By looking at his 10,000-meter times Daniels figured

123

Salazar could run 2:07:51. Salazar covered the 26.2-mile distance in 2:08:13 at the New York Marathon that year, breaking Derek Clayton's record by 20 seconds. So Daniels was off a bit.

"I'm often a little lower than the actual time," he explains. "It's impossible to consider every factor. For one thing, when I estimate road races from track times, I have to add roughly two minutes to account for the differences in terrain. But other unquantifiable factors like heat, wind, the caliber of the competition, and crowds also come into play.

"People have a fondness for round numbers. Right now everyone's wondering when and if women will break the 2:20 mark in the marathon. [The current women's world record, set by Joan Benoit at the 1983 Boston marathon, is 2:22:42.] I think they will, but records progress together. When the women's 10,000-meter bests improve to under 31 minutes, you'll see a marathon mark in the low 2:20's."

Raisa Sadreydinova of the USSR holds the 10,000-meter record of 31:35.3, so the women's track performances appear to be right on schedule. Most experts agree that women's records will change more dramatically than men's. During the last twelve years, for instance, the women's record for the marathon has dropped more than twenty-three minutes, while the men's mark dropped only twenty-two seconds. "Men have run the distance thousands of times," says Daniels. "In a sense, women are just getting started."

For that reason the men's records are generally much closer to the limits of human performance. The world record for the 100-meter sprint, set at 9.95 by Jim Hines during the 1968 Olympics, was only recently broken by Calvin Smith. "The human body hasn't evolved much in the past one hundred years," says biomechanicist Gideon Ariel. "All the changes we're seeing today come as a result of improved technology and training." In one case Ariel analyzed the performances of Hines and one of the other top sprinters of the twentieth century, Jesse Owens. He examined films of Owens at the 1936 Olympics and Hines at the Olympic Trials in 1968 and simulated a race between the two under identical conditions. They tied in his experiment, both matching the world record. In 1936 Owen's time was 10.3.

"Owens lost speed because his foot slipped slightly on the track, he didn't have any starting blocks, and the track didn't have the rebounding characteristics of today's surfaces," explains Ariel.

The change from cinders to synthetic rubber tracks and the development of lighter running shoes has had a big effect on records, subtracting, some experts believe, a second per lap from times of twenty years ago. And in the pole vault the world record shot up nine inches the year the fiber glass pole was introduced; it had taken the previous twenty years to raise the record that same nine inches.

New techniques can have a similar effect. In high jumping Richard Fosbury's backward twist over the bar led to a two-and-one-half-inch jump in the world record. And as the athletes go higher and faster, the strategy in the events themselves change. The mile used to be a "hold back until the last lap" sort of race, but now it's almost a full-out sprint from the start. Pacing by "rabbits" (fast runners who drop out after the first few laps) is also common. Runners today rely more on their split times—single lap times, for example—to calculate their reserves as they compete. Alberto Salazar followed a truck bearing a digital display of his time through the streets of New York en route to his 1981 world record.

But while athletes may not have reached the limits of performance, they may have come close to reaching the limits of training. "There is a limit to what the body can take," says Bob Hersh of the National Records Committee. "If a runner racks up 200 miles per week, chances are he'll break down. Even the workhorses only put in 140 miles these days. And it's not hard for a world-class athlete to run eight-minute miles for four hours a day."

Finding out just how much training athletes can take is increasingly becoming part of their training. They spend a lot of time now on treadmills or giving blood to help their trainers learn what's going on inside them. Runners even have their rate of oxygen consumption measured and their muscles fiber-typed, in part to insure that the event they're participating in is the best one for their natural abilities.

Daniels believes that how much oxygen an athlete can take in during competition is the best indicator of his or her innate efficiency and potential. "You can look at someone who's five feet two inches and 135 pounds, and know that he shouldn't become a shot-putter, just as someone who is six feet six inches shouldn't aspire to be a gymnast or a jockey. But if you look at a person who is five feet ten inches and 165 pounds you'd be hard pressed to tell whether he's a potential record holder. But if you test him for his VO_2 max [the amount of oxygen taken in to fuel the muscles during exertion], you'll find out right away how he'll stack up." VO_2 max varies from individual to individual. Marathon world-record holder Benoit has a VO_2 max comparable to that of many world-class male athletes. Most men use about seventy millimeters of oxygen per kilogram of body weight per minute; Benoit uses seventy-nine. If Daniels is right, Benoit has even better performances in her, perhaps even that sub–2:20 marathon.

A few years ago, muscle fiber–typing was the rage of the sports medicine community. Put a small sample of muscle from a well-exercised part of the body (like the muscle at the back of the thigh) under the microscope and it will reveal the percentage of fast and slow twitch muscle cells it contains. Fast twitch fibers are adapted for quick, short bursts of speed, while slow twitch muscles contract regularly over long periods of time and are used primarily in endurance events.

Basically, slow twitch fibers contain more mitochondria—organelles that use oxygen to produce energy—than fast twitch fibers and therefore are useful in long distance events. But fast twitch fibers contract quickly and with more force. A person can train for speed or endurance and improve tremendously, but he will be at a disadvantage when competing with someone who has the better genetic muscle makeup. Most people have roughly equal portions of slow and fast twitch muscles, falling somewhere between the 85 percent fast twitch muscles of sprinters like Carl Lewis and the 92 percent slow twitch muscles of marathoner Salazar.

But even the right muscle fiber type and VO_2 max doesn't guarantee a world record. "Most of today's elite distance runners have more than 80 percent slow twitch fibers and train similarly," says physiologist Lawrence Armstrong of Ball State's Human Performance Laboratory. "So it's the subtle things like psychology, personality types, and coaching that separate the champions from the rest of the crowd. In a sense, the champions have to have a great capacity for punishing themselves, for running through fatigue."

It is possible, of course, that very few of the world records in the next century will resemble those of today. Certainly those of the 1880s don't look much like the current marks. Back then the four-minute mile was an unreached goal. Now it is possible that a 3:30 mile may be achieved. That would require milers to maintain a speed roughly equivalent to today's 800-meter runners (around seventeen miles per hour), but then as physiologist Thomas Fahey remarks, "As long as there are records to break, athletes will find some way to do it."

"A world record is an awesome achievement," says Bob Hersh. "I have the utmost respect for a person who does something faster, farther, and better than any of the millions of people who have tried it before. And it's quantifiable. You can't say 'this is the best work of literature' or 'this is the best operation that's ever been performed.' But you can say that Sebastian Coe ran the mile faster than anyone else ever did."

At least until someone borrows it from him.

Part IV

FORM

Ballistics of Speed Skiing

ERIC PERLMAN

His body tucked tight as a bullet, world-champion speed skier Steve McKinney hurtled down the mountainside. Red, blue, and green marker flags flashed past as dark blurs against the snow. There were no turns, no banks, no deviations from the path straight down the forty-degree slope. For McKinney, there was no time for fear or doubt. There was only the furious rush of speed and the quest for the perfect ballistic posture.

At Kilometro Lanzado, "the flying kilometer," 3,500 meters high in the Chilean Andes at Portillo, pure speed is the goal. Every year a host of international daredevil skiers comes gunning for the unpowered land-speed record.

During the more than twenty years that speed skiing competition has existed, 200 kilometers per hour seemed the unreachable upper limit. Like the runner's four-minute mile, it eluded the most determined effort of the world's fastest skiers. Some said that 200 kilometers per hour was impossible; it stood on the far side of "terminal velocity," a speed moving bodies cannot exceed.

But terminal velocity is a variable, dependent on the density of the air and the shape, mass, and surface area of the falling body. The terminal velocity of a mist particle is zero when its tiny mass hangs suspended in dense cloud. Peregrine falcons dive for prey at estimated speeds of more than 280 kilometers per hour. Talons extended like a speed skier's fists, wings swept back to minimize

Hurtling down Kilometro Lanzado, Steve McKinney attempts to become the world's first skier to break the "impossible" 200 kilometer-per-hour barrier (124 miles per hour). (© Bob Woodall)

turbulence and drag, an attacking falcon comes close to achieving the perfect ballistic posture.

Near sea level, the terminal velocity of an aerodynamically positioned sky diver in free fall is about 300 kilometers per hour. In 1960 sky diver Captain Joseph Kittinger, USAF, wearing a pressurized suit, leaped out of a high-altitude balloon at 31,000 meters, where the air is very thin. He attained a supersonic speed of more than 900 kilometers per hour at an elevation of approximately 28,000 meters, then began to slow down as atmospheric density increased.

Though they race in the comparatively dense air at 3,500 meters, speed skiers have a long way to go before they reach their true ultimate speed. Kilometro Lanzado and its counterpart, Kilometro Lanciato in Cerbinia, Italy, are the proving grounds for the latest innovations in ski-manufacturing technology. Every piece of

McKinney's equipment, every bend of his body was designed to minimize wind resistance and maximize speed.

As McKinney accelerated down the course, he crouched low. His hair was tucked inside an arrowhead-shaped helmet; his slick, skintight nylon suit could have been painted on. The ski poles were bent to wrap around his waist and jut out straight behind his back. Held close to his body, the poles stabilized his outstretched fists as they broke the mounting wall of wind. Where they thrust behind him, the spheroid ski poles' baskets and protruding tips gave directional stability, like the fins of a missile or an arrow's fletching. His 235-centimeter skis with the low-cut, flat-front shovels shuddered and sang as they alternated between riding the surface of the snow and a river of air, two to five centimeters off the ground.

The wind resistance McKinney fought so hard to penetrate takes two forms. The first, called form drag, arises when the pattern of airflow around a moving body does not close up at the tail but separates to form a turbulent wake. The difference in pressures between the nose and the tail sucks the moving body backward into the lower pressure of the turbulence behind it. A properly streamlined body, like a jet fighter's or a dolphin's, nearly eliminates this kind of resistance because the shape of the tail matches the contour lines of the airflows or water flows that it generates at typical operating speed.

The second kind of resistance, called skin friction, is caused by air sticking to the surface of a moving body and forming an adhesive boundary layer. When a body moves through air, the air molecules immediately contacting the body adhere and move at the same speed. Though this adhesive boundary layer of air is usually quite thin, it varies with the shape, size, and speed of the moving object. More than half the horsepower used by aircraft cruising at high speeds is required to overcome the resistance of skin friction.

McKinney trimmed his tuck position, knees to thorax, the better to penetrate the wall of wind. The speed climbed: 180 . . . 190 . . . 195 kilometers per hour. Mind blank, attention honed, reflexes ready to fire and fight for microfine adjustments, McKinney powered through the speed trap with the rushing roar of a jet.

Two pairs of photocells marked his entry and another two pairs flashed at his exit, one hundred meters later. A computer worked out the average speed to a thousandth of a kilometer per hour. And then the announcement came: "Steve McKinney, USA, 200.22 kilometers per hour, a new world's record"—October 1, 1978.

Skiing on Air

SUE HOOVER EPSTEIN

Ski jumping has an image problem in the United States, due in part to a hapless Yugoslavian who careened down a ramp twelve years ago, lost his balance, and crashed in a one-man avalanche of skis and snow. Each week, millions of viewers still watch Vienko Bogatej's spectacular fall as part of the introduction to ABC's "Wide World of Sports."

"People think of jumpers as daredevils," says Jim Page, Nordic director of the U.S. Ski Team. "Or that we must be on drugs or demented to even try it. But really it's no more dangerous than any other sport."

A typical flight on a ninety-meter ski jump lasts about four seconds, says Page, "so in a sense whole careers can be measured in minutes." U.S. Olympic coach Greg Windsperger wanted his top jumpers to take seven hundred training jumps before the start of the 1984 Olympics in Sarajevo, Yugoslavia, giving them about an hour of air time.

Many U.S. jumpers come out of the small rural communities near the twenty ski jumps longer than seventy meters—places like Ishpeming, Michigan, Brattleboro, Vermont, and Lake Placid, New York. Still, the best U.S. jumpers spend most of their time out of town. Until this year the U.S. Ski Team had to practice overseas during the summer. The Europeans have used plastic mats for off-season training since the 1960s, but only in July 1983 did the Olympic Training Center in Lake Placid install the shiny, looped mats on seventy-, forty-, and fifteen-meter hills there. The mats, which resemble spaghetti, are bound together and spread on the takeoff and landing areas. The hills, looking like oversized thatched roofs, are then sprayed with water to keep them slick.

The skiers reach roughly the same distances on plastic as they do on the real thing. The plastic may also be slightly safer than snow, says Steve Gaskill, a former national coach who currently works with young jumpers. "The skiers don't dig in when they fall like they do in snow. I hear it can burn a little, though."

Even with their own mats, American jumpers continue to train with Europeans. "It's good for them to gauge themselves against the best," says Gaskill. "Then they can see what speeds and styles are winning competitions."

In the 1860s Norwegian cross-country skiers often amused themselves by hopping off small hills. But the sport of ski jumping didn't really take off until Sondre Norheim of Telemark, Norway, created the now-classic down-on-one-knee landing. Moving the right knee forward keeps the jumper's balance centered and transfers the shock of landing to the hips and thighs. With the telemark landing, distances increased from the 30.9 meters of Norheim's first officially measured jump to the current record of 180 meters.

The Finns and Germans were the first to change their upright jumping profiles to a forward lean and promptly started taking titles from the Norwegians. Early film studies of the flying Finns by Swiss engineer Reinhard Straumann and Norwegian coach Thorleif Schjelderup revealed that the most successful jumpers were making their bodies into airfoils. Leaning forward from their ankles at about a thirty-degree angle to their skis, jumpers created lift as the air flowed faster over the curve of their bodies.

But before a skier can fly, he has to take off. In a crouch, the skier must keep his weight centered over his skis as he picks up speed down the ramp. At the takeoff, the critical part of the jump, he thrusts himself out over his skis and straightens his legs to bring his ski tips up. A jumper pushes his hips forward at takeoff, actually generating from three to five miles per hour of extra speed from this motion.

Once aloft, a jumper must keep his upper body at a constant angle to his skis. "It's very difficult to prevent further extension," says Gaskill. "In basketball and many other sports you can extend

Once aloft, a jumper must keep the tips of his skis angled into the wind and his back curved—forming an airfoil not unlike that of an airplane wing. (© Hubert Schrieb/FOS)

your body fully, but in ski jumping you have to maintain a certain angle and stay flat so you don't lose speed. It's close to diving and gymnastics in this respect."

Because the longest jump is almost by definition the most aerodynamically sound, it usually wins the competition. Jumpers can lose points, however, by wobbling in the air, for instance, or letting a hand touch the snow on landing.

According to five-time Olympian Art Devlin, early jumpers had a lot more to worry about than form. "It's much safer now," says Devlin. "Before the 1950s each community with a ski hill built it differently. One of the biggest problems we had was knowing how fast we were going and where we were going to land." Devlin speaks from experience, having once outjumped a hill in Leavenworth, Washington, by forty feet, tearing the cartilage in his right knee and missing his chance to compete in the 1958 Olympics.

Now all ski hills conform to the same engineering specs and

have movable starting gates. At ski jumping competitions, young noncompetitors called forejumpers test snow conditions and inrun speeds—usually about fifty-five miles per hour on a seventy-meter hill—before the starting gate is set. Wind and snow conditions can change during an event, though, so the judges watch for unusual jumps that might indicate that the inrun has iced up or the wind has gusted out of the safe thirty-five-miles-per-hour range.

Judging made a big difference in the outcome of the seventy-meter event at the 1980 Lake Placid Olympics. Jeff Davis, an American whose best jump ever had been seventy-nine and a half meters, made a leap of ninety-one meters, well past the sixteen-meter-long landing area. His was the ninth jump of the event. After a delay the competition jury ruled the jumpers were going too fast on the ramp, and moved the starting gate down, forcing everyone to jump again.

Davis made two creditable leaps of eighty and eighty-four meters to finish seventeenth, and the competition was won by Austrian Anton Innauer with jumps of eighty-eight and ninety meters. Davis handled his disappointment with grace, but others felt that if someone with an international reputation had jumped ninety-one meters it would have stood. The judges were concerned that if Davis flew so far, the top jumpers would surely sail into hazardous territory beyond the hill's landing area.

The athletes themselves just want to outjump the competition, landing zone or no landing zone. Some try out unusual styles to get greater distance, risking the disapprobation of style judges. In the old days everything from flapping the arms to bowing from the waist was seen. More recently, a style called the delta-wing came into vogue. Most jumpers keep their skis parellel about six inches apart, but others have been able to successfully complete jumps with their skis in a snowplow or triangular formation. Canada's leading jumper, Steve Collins, used this technique for some time but has since abandoned it.

"Collins did the delta-wing naturally," says Jim Page. "He's a great wind skier. His size (roughly 105 pounds) may have something to do with it. He floats on the wind—doesn't drill through it like other skiers."

According to Page, floaters get the most out of every flight, riding the wind and maintaining lift as long as possible. Power jumpers acccelerate more at takeoff but often do not have the best aerodynamic posture getting down the hill. Floaters tend to adjust better to ideal days with uphill breezes when they can take advantage of the lift.

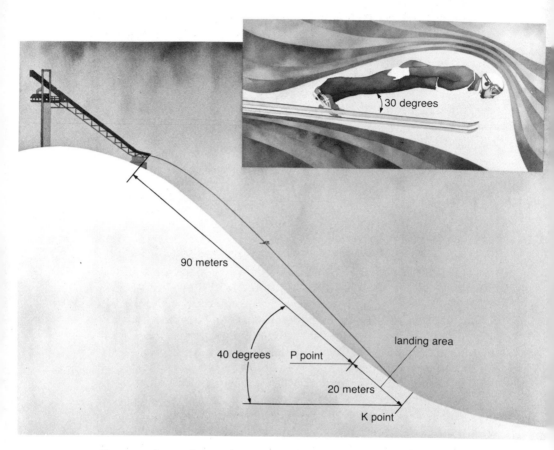

30 degrees

90 meters

40 degrees   P point

landing area

20 meters

K point

Shooting down the ramp of a ninety-meter ski jump at sixty miles per hour, jumpers launch themselves off a lip angled at eleven degrees below horizontal. Though they fly more than ninety meters, they rarely get more than six meters off the ground. This jump at the 1984 Olympic Winter Games at Sarajevo, Yugoslavia, has a "safe landing" area—designated on ski jumps as the part of the hill between the "P" and "K" points—that is twenty meters. Past the K point the flattened slope makes landing dangerous. (© Greg Harlin/Stansbury, Ronsaville, Wood, Inc.)

To prolong the four seconds they have to find their best inflight posture, U.S. skiers have been hanging around a wind tunnel at the Arvin-Calspan Advanced Technology Center in Buffalo, New York. Michael Holden, an aeronautical engineer at Calspan who works primarily on space shuttle reentry speeds, puts jumpers in a harness attached by cables to the ceiling. He measures lift through changes in the vertical tension and drag through changes in the horizontal tension. Jumpers learn to cut their drag, for example, by repositioning their hands and heads or by adjusting the angle of their skis.

"The jumpers go off the inrun around fifty five miles per hour like cannonballs," explains Holden. "They don't have much time to think. But they feel as if they're flying, and if they can get themselves in the right position after takeoff, they can have a longer ride. The key is to find a good position and work it, rotating forward to maintain the angle and speed."

Some people, Art Devlin among them, feel that all this technology takes away from the adventuresome spirit of the sport. "Why make the sport so complicated that you need to be an aeronautical engineer to do it?"

Perhaps because today, as Mike Holden puts it, "In ski jumping the difference between winning and losing is flying."

Propellers, Paddlewheels, and Swimming Faster

JOHN JEROME

When the 1976 Montreal Olympics came to an end, there was only one world record in swimming left on the books that was more than two months old. Mark Spitz' seven gold-medal times at Munich in 1972 wouldn't even have qualified him for the 1984 U.S. team tryouts. This remarkable transience of swimming records over the past ten years cannot be attributed to a new diet or a wonder drug or a spurt in the sport's popularity. It's simpler than that: Swimmers and their coaches have learned how to better use the hands, arms, and feet so that the body moves through the water more like a motorboat than a canoe.

It is true that some of the improvements in swimming performance have come from attention to external details. Water itself hasn't gotten any faster, but swimming pools definitely have. It is turbulence that most interferes with fast swimming—choppy water on the surface, roiling minicurrents below—and when racers dive in eight abreast or hit the wall with flip turns virtually in unison, you get aquatic chaos. The new pools hold the water still. They are at least seven and a half feet deep to reduce rebound off the bottom; at the edges, contoured gutters swallow waves and give no turbulence back into the pool. Even the dividers between the racing lanes are strung with intricately finned disks to damp out wavelets. In a fast pool, every swimmer in every lane gets to race in quiet water.

Swimsuits themselves are now made from exotic stretch nylons and lycras—thinner, tighter, and much slicker—to reduce drag. One new fabric is polymer-coated so the surface repels water; swimsuits made from it must be put on wet so you won't split the polymer coating and so it seals around the contours of your body. (The suit, of course, is not recommended for distances of more than 400 meters because perspiration under it can cause extra drag.) Coaches also note that the pace of record breaking accelerated with the introduction of racing goggles. Goggles not only improve vision, but by reducing eye irritation from pool chemicals, allow swimmers to swim longer, permitting two or three times more training yardage per workout. And to swimmers, yardage is like money in the bank.

But it is in the biomechanics and the biochemistry of swimming that subtler and more profound discoveries are being made. It turns out that what a good swimmer does in the water is quite different not only from what he appears to be doing but usually from what he thinks he's doing.

For example, in strokes where the arms move forward above the water—the freestyle, backstroke, and butterfly—a good swimmer's hand may come out of the water ahead of the point where it entered. "Basically, what the swimmer is doing is anchoring the hand in the water, then pulling the body past it," says Ernest Maglischo of Bakersfield State College in California. "When the hand first enters the water, it slips forward for quite a distance, so by the time it comes out, it's still ahead of where it went in."

Swimmers themselves have trouble believing this. The strongest impression they have is that they're pulling their hand through the water like a canoe paddle. In particular, inexperienced swimmers try to swim this way, pulling the arm and hand in a straight line beneath the body. This is not only ineffective but difficult to do: Water pressure makes the hand and arm dart from side to side, and it takes a great deal of muscle to overcome this darting motion. When powerful swimmers pull hard, the hands seem naturally to pursue an S-shaped path. Until the mid-1960s, this S-curve was regarded as a technical flaw, and world-class racers struggled to erase it from their strokes. But it kept winning races.

Maglischo, a swim coach as well as an exercise physiologist, has shown why. The hand and forearm, he says, are used less as paddles than as propeller blades. A paddle pulls straight back, its blade at right angles to a boat's motion. The boat is pushed forward because water resists the movement of the blade. Propellers, on the other hand, are pitched to thrust water back, producing a lift that pushes the boat forward—a much more efficient means of propulsion. In the swimming stroke, the hand and the forearm form a plane. As this plane slides through the water, the fluid is divided into two streams. The stream that goes over the curved back of the hand (which resembles the front of a propeller blade) must travel faster than the stream over the palm of the hand (which resembles the back of the blade). This creates less pressure on the top of the hand and more pressure on the palm, generating a force akin to the lift generated by an airplane wing.

"I give a little demonstration to my swimmers at the beginning of a season," says Maglischo. "I put a toy paddle boat and propeller boat in the water. The propeller boat beats the hell out of the paddle boat every time."

Thus the search for swimming efficiency is a search for the angles of attack of the hand, arms, and even the feet that will provide the greatest forward pressure for the energy put into them. Since we can't rotate the hands and feet like propeller blades, we are forced to find the pitch in other ways. The S-curve is the best other way.

Imagine a line through the water, dead ahead and parallel to the surface, representing the perfect line the swimmer wants to go. As Maglischo describes a good swimming stroke, the hand follows an S-shaped path when viewed both from below and from the side. The hand slides downward and forward at entry. It then sweeps slightly outward—angled away from the body, fingers pointing toward the bottom of the pool—to a point a little outside the shoulder. Next it sweeps back in under the body toward the swimmer's stomach, the elbow bending at about ninety degrees. It finishes with another swirl to the outside again, near the hip, pitched outward and backward. Out, in, out again. "Actually," Maglischo says, "you have to describe it three-dimensionally to make sense of it. There's an in-and-out S-curve combined with an up-and-down S-curve."

In sweeping the hand through its S, the swimmer's cue that he or she is getting the most effective pull is the resistance of the water to the hand. But sometimes the most resistance comes when the pull is misaligned, dragging or pushing the body inefficiently off-line. In fact, lift propulsion is usually a few degrees off the direction a swimmer is pulling. The swimmer must learn to sense the tiniest deflections from that ideal straight line forward and adjust the pitch of the hand. You pull at whatever angle you can put the most muscle behind; you pitch the hand and arm so the force that's created pulls you straight ahead. In this way a swimmer gets the best possible speed for a given output of energy.

A mechanically perfect stroke without the strength and endurance to keep the arms stroking won't win any races either. That's why Maglischo and exercise physiologist David Costill, director of the Human Performance Lab at Ball State University, are not just interested in how to swim, but in how much. It is a question that has always been answered with the theory that more just had to be better.

Top-level swimmers are gluttons for work, regularly putting in six to nine miles of training per day. Swimming burns energy at about four times the rate of running; an eight-mile day is roughly equivalent to a thirty-two-mile run (and even marathon runners seldom train much over twenty miles a day). This conversion is not perfectly linear, of course, and some advantages go to the swimmer: The water is an automatic cooling system, and there's no wear and tear on the body from hitting the ground.

All athletic training can be placed somewhere on a scale of 0 to 100 percent, whether the percentage is of maximum speed, effort, distance, or any other training value. All else being equal, the athlete who trains hardest—closest to 100 percent in all measures —will perform best. Unfortunately, training is stress, and when stress gets too high, the body stops improving and starts breaking down. Runners who overtrain eventually injure their ankles or knees or shins; when swimmers train themselves into debilitation, they tend to suffer from depression, insomnia, and overexhaustion, which in turn may lead to a rash of physical symptoms that mimic flu and can turn into something very much like mononucleosis. The goal is to find where, on that 0-to-100 percent scale, the maximum gain can be obtained—without breakdown.

By looking at swimmers training at different distances and speeds, Costill is trying to quantify where on the scale a swimmer is training. For example, he corrals tired swimmers during and after practice and measures their levels of lactic acid and blood pH. Lactic acid and low pH inhibit muscle action, and thus their levels indicate how long it will take for the muscle to be able to perform at top speed again. Or he checks the levels of glycogen—the muscle fuel of long distance swimming—for signs of depletion. From this research Costill has generated a "Training Intensity" rating. He figures that training intensities below eighty on his scale can be repeated several days in a row without the swimmer's beginning to break down. A workout with a training intensity of eighty or above should be followed by a day off—or at the very least, a workout of much less intensity to permit recovery.

He also can read a swimmer's "swim-rest ratio"—an index of how aerobic or anaerobic a given set has been (the higher the ratio the more aerobic it is). Swimming races of more than two hundred meters in length are mostly aerobic in nature, the muscles getting the bulk of their energy from consumption of oxygen. The majority of swimming races are 200 meters or shorter, however, and are highly anaerobic, powered primarily by "on-board" energy supplies in the muscles and blood, building up an oxygen debt that is paid back only after the event is over. "Most exercise research in recent years has been with aerobic metabolism," Costill says. "But most of swim racing is highly anaerobic. There's a lot we don't yet know about the anaerobic side of things."

Physics on the High Dive

PATRICK COOKE

Generally speaking, people do not care much for having the planet pulled out from under them. They learn to step lightly around tumbling mats and back away in the presence of an ordinary trampoline. Even the high dive at the local pool is a plank whose path seems better left unwalked.

No less intimidating is the physics behind the flips and twists that spring forth from such places. "Angular momentum isn't something people readily understand," says physicist Cliff Frolich of the University of Texas, not that understanding this principle of physics will make any ground-grabber the next Nadia Comenich. Frolich, once a collegiate diver, has for several years been studying the effects of angular momentum, the amount of spin an object generates as it moves. "It's something we use daily in our lives," he says. It's also seen in any of those sports where the human body twirls like a pinwheel—gymnastics, diving, skating, ski ballet, even circus acrobatics.

Anything that rotates has angular momentum, and the amount depends on the object's rotational speed, weight, and how the weight is distributed. From toy tops and Frisbees to gyroscopes and solar systems, nothing is too large or too small to spin—and keep spinning. Mass of all kinds resists being put into motion, but such objects are just as determined to keep moving once they start and will spin on until interfered with by an outside force like

This forward two-and-a-half somersault with two twists, above, shows that divers can initiate twists in mid-air. The diver begins with a straight somersault, then throws one arm up and one down to turn his body slightly away from a forward somersaulting alignment. Because the diver's angular momentum is conserved, other parts of his body rotate and the diver begins to twist. If the diver were able to throw his arms hard enough to realign his body so that it was perpendicular to the somersaulting axis all his angular momentum would go into twisting; since he is somewhere between the two axes, he twists and somersaults. (© Greg Hargreaves/Hellman Design Assoc.)

friction or gravity. Lifting the whirling front wheel of a bicycle off the ground, for example, gives a good idea of the power of angular momentum. The faster it spins, the harder it becomes to turn the wheel and force it right or left of its original direction.

Somersaults and twists are the two basic rotational motions around which most spinning athletes build their routines. The first difficult step is getting the body weight launched into the air. In most cases that requires nothing more than a mighty push from the legs, but a trampoline or a diving board helps.

When a diver begins a somersaulting dive, for instance, the action of the board propels him upward and outward. The torques applied through the board to the diver's body are so strong that rotation begins regardless of the position he assumes. Beginners, in fact, sometimes painfully discover that their legs overtake their heads even during a simple swan dive. It's angular momentum that lands them on their backs. For most dives a skilled diver "throws" or rotates his arms forward (or backward) in the direction of the somersault at takeoff.

The speed at which a diver then somersaults depends on the positioning of his body. In a layout position, for example, where the body is stretched lengthwise, he will somersault slowly because his weight is far from the axis of rotation. Last year, 130-pound Miguel Vazquez made circus history by doing just the opposite. By lowering his head and raising his knees, he put his body into what is known as a tuck posture. With all his weight snuggled around his center of gravity, he was compact enough to complete four rapid back somersaults from the trapeze before passing into the trusting arms of a partner.

The acrobatics become less straightforward when the athletes start adding twists to their routines. For years it was believed that all twists had to begin while an athlete was still in contact with the ground. (These are indeed done, and are called torque twists—the same type of movement a skater uses to jump from the ice and turn in the air.) But Frolich, who has studied high-speed films of divers and trampolinists, says that as long as there is some angular momentum in any direction, it can be used to change direction in mid-air. A tumbler can begin twisting, he says, even if he does not initiate it until well after he has left the ground.

Though the effects of simultaneous spinning and twisting dizzies the crowd, the actual body moves an athlete makes in adding his twist are surprisingly slight. For example, if an athlete somersaulting forward in a layout position throws an arm up over his head and the other down to his waist, he suddenly knocks himself away from his somersaulting axis and diminishes his somersaulting motion. Because angular momentum must remain constant while a tumbler is airborne, however, the body compensates for the loss of forward spin by beginning to twist on a vertical axis, which runs from head to toe. No tumbler is able to throw hard enough to convert all somersaulting to twisting, so the body does both—at a rate of about three twists per somersault because weight is more closely packed along the vertical axis.

The only problem now, as Frolich points out, is that "gravity makes you hurry up." Naturally, the more altitude, the more time for gyration; but even if a diver can get ten feet above the board he has no more than 1.5 seconds to complete a routine. His final airborne act is to stop the twist and once again realign with his somersaulting axis, which he does by quickly rethrowing his arms for what he hopes will be a graceful descent.

"You have to use all your senses when diving," says Greg Louganis, winner of more than twenty U.S. national titles and generally considered the greatest diver in the history of the sport. "Visual cues are important when you're spinning. I do a reverse three-and-a-half, and I know I'll see the water three times. Also, you listen to the sound of the board. When I leave it I know just when I am going to hit the water."

Another type of turn used by some athletes that intrigues scientists is the zero angular momentum, or cat twist. Experts have long disagreed about why a cat lands on its feet when dropped from an upside-down position. What starts it turning? Frolich believes that this is actually an angular momentum maneuver that even humans can do. For instance, an astronaut in space can create momentum by throwing an arm across the chest to turn the upper body. Because momentum is conserved, the hips must swing around in the other direction, in effect canceling out the momentum that had been generated. Rotation then stops, and other maneuvers have to be made to reach desired position. It is because of this type of cat twisting, for instance, that trampoline tumblers can perform moves like the "swivel hips," where an athlete can change direction not only on the way up but on the way down as well.

All this is fine for those of us safely anchored to the bleachers, but how much knowledge of mechanics actually helps an athlete pull a better stunt? "Not much," says Louganis. "I just get up there and do it. I know what I have to do and what it should feel like. Timing is really the most important thing." Ron O'Brien, current coach of Louganis and other Olympic hopefuls, agrees:

"We train for concentration, but it's impossible to keep everything in mind." His athletes, he says, "have to have an understanding of the mechanics involved, but what's critical is an awareness of the movement. Sometimes the less you think, the better off you are."

Pole Vaulting: Biomechanics at the Bar

JOHN JEROME

There's no great mystery about the mechanics of the pole vault. You accelerate your body to maximum speed, store the resulting momentum briefly in a fiber glass pole, convert the momentum from horizontal to vertical motion and add to it whatever additional muscular force you can muster, and direct this new motion upward to clear the crossbar. Zip down the runway, slam the pole into the box, bend the pole into a great circular bow: Up and over you go.

The image is of flowing activity, if not exactly grace, but that doesn't make it simple. A good vaulter must be as fast as a sprinter, as agile as a gymnast, as strong as a weight lifter, and as brave as a cliff diver—and he must be able to link these attributes as he links the serial steps in his event.

Sometime this summer of '84 the world record in the pole vault is expected to spurt upward from last season's nineteen feet and three-quarters inches to, perhaps, nineteen feet six inches; track buffs are talking confidently about twenty feet. Smart money is on Billy Olson, a cherubic twenty-four-year-old athlete from Abilene, Texas, to do much of the record breaking. But Olson, the only vaulter to clear nineteen feet indoors, is merely the leader of a renaissance in American vaulting. And although there are young vaulters developing to world-class levels elsewhere in the United States—notably Dave Volz of Indiana and Joe Dial of Oklahoma

State—the hotbed of the sport is Abilene Christian University, in Abilene, Texas. Olson, a graduate student, trains there, and for company he has three eighteen-foot vaulters on the team. Abilene Christian University's track coach, Don Hood, is the reason why. He's a crafty coach and an inspired teacher of the mechanics of track and field.

"To start with," Hood says, "you can't have a great vault without a great run. If you're using all your speed and hitting your takeoff point precisely, you can be relatively assured of a good vault. I can turn my back and hear if it's a good run. If the vaulter is tentative, he'll slow down on the last three steps. You'll hear his footsteps go pow pow pow . . . pow . . . pow. If it's a good run, he'll go pow pow powpow POW."

A tentative vaulter, he says, is likely to take off with his foot ahead of his upper hand on the pole, leaning backward and thereby reducing the arc through which he can swing his body weight to generate upward force. Bad physics. The tentative vaulter also wants to hug that pole, holding it right by his ear instead of with his arms extended straight. When the arms are extended the pole is at a greater angle to the ground and will start to bend more easily. If he doesn't get the bend started properly, he'll end up absorbing the force of the run with his arms instead of transferring it to the pole. More bad physics.

The next step, virtually simultaneous, requires the vaulter to jump vigorously—not up but out, like a long jumper, to preserve his momentum against the shock of the pole-plant. The vaulting pole is a very slender diving board or trampoline. To get the pole to bend as far as possible the vaulter must delay swinging forward as long as he can. "If you rock back too early," Hood says, "you're going to die at the top. Then when you try to pull up, turn, and push off to clear the bar, you don't have any momentum. You're trying to lift dead weight over the bar. By pulling the knees into the chest, you're getting your weight closer to the point of support, which is your grip on the pole. Then when you pull, you're going to go up. If your legs are still straight—or if you're piked at the waist—you're going to go out instead of up when you pull."

As the force generated by the spring of the pole begins to wane, the vaulter pulls and vigorously shoots his feet upward. "He can add four to six inches to the vault that way," Hood says. "Then for the pull and the turn we want the vaulter to stay as close to the pole as possible because he can pull a lot more weight that way. That means he has to have a good relationship to the pole. Underwater vaulting is good for teaching that."

Hood has developed underwater vaulting into a powerful teaching tool. Wearing a diving mask and holding his breath, a vaulter sits on the bottom of the deep end of a swimming pool with a shortened pole vertically beside him. He reaches up and gets a proper handhold, then begins his rock-back, tucks, pulls, shoots his legs, turns, and pushes off to clear a "crossbar"—a submerged garden hose. "Underwater vaulting helps you reinforce the picture you have in your mind," Hood says. After an underwater vaulting session, one of Hood's vaulters raised his best effort by more than seven inches the next day. "You can get a lot bigger pushoff underwater. The water magnifies everything, slows everythng down, and allows you to go through the whole motion without worrying about falling backward."

Fear of falling backward is one of the vaulter's biggest problems, according to Hood. "If a vaulter is afraid, as soon as he leaves the ground he wants to pull himself up on the pole. There's no way he can rock back if he does that, no way he can get his hips up. I remember when I was vaulting, the coach would tell me to do something, and the moment I left the ground my mind would go blank. I couldn't think in the air because I wasn't ever in the air enough. When fear takes over, the thought processes just cease."

Fear plays an almost technical role in pole vaulting, because, as Billy Olson says, you can't get on a big pole if you're scared. ("Big" refers to stiffness rather than length.) And you're not going to jump very high if you don't get on a big pole.

Vaulting poles are measured in meters. "Most of the world-class vaulters jump on five-meter poles," Olson says. "That's roughly sixteen feet five inches. They make poles as long as five-meters-twenty—that's seventeen feet four and three-quarters inches— but not many people jump on poles that long." Nobody vaulting now is fast enough and strong enough to put the additional length to use.

The stiffness of the pole is measured by deflection. The poles are suspended on supports fifteen feet nine inches apart, and a fifty-pound weight is hung at the midpoint. The deflection is measured in centimeters. "My poles run from 15.6 to 14.5 centimeters in deflection," says Olson. "A 15.6 is softer than a 14.5."

Vaulting poles are made from three pieces of fiber glass. The first layer is tape wrapped in a spiral around an aluminum template, then reversed for a second layer. The spiral tape gives the pole strength around its circumference to keep it from "kinking." Then a rectangular piece as long as the pole is wrapped around it several times. The fibers in this layer are woven so that 97 percent of them will run lengthwise down the pole, giving it its trampoline strength. The outermost layer is called the sail piece. It's cut a little like an equilateral triangle with one corner lopped off. The sail piece is wrapped from the base so that less fiber glass is laid on the top and bottom of the pole than in the middle as it is wound. By adjusting the cut of the sail piece the manufacturers can adjust the stiffness of the pole—the more fiber glass in the pole, the harder it is to bend and the more energy it returns. Finally, the pole is heated in a pressurized oven at 325 degrees and the aluminum template is removed. Because the template is slightly tapered—it is fifty thousandths of an inch thicker at its base to make its removal easier—all poles are also slightly thicker at one end.

In a single year—back in 1963—the world record for the pole vault was revised upward by roughly nine inches, thanks to the fiber glass pole. With vaulters jumping on bamboo, steel, or aluminum poles, the previous nine-inch rise in the record had taken twenty years; in the next twenty years it rose nearly three feet. This might suggest that another technological change in the pole is likely to revise the sport all over again. Billy Olson doesn't think so. "They've improved the poles a lot in the last five years—they made them lighter and smaller in diameter—but they've made poles out of just about everything they can think of, and the one we've got now is the one that works best."

Good vaulters get to know their poles extremely well. "I usually carry three to five poles to a meet," Olson says. "I take along one pole that's a little small [soft] for me. If conditions weren't right or I was hurting or something, then I could still jump with that one. The rest are almost exactly the same."

Olson's advantages as a vaulter are great speed and great upper body strength. "A slower vaulter might want the same bend that I have, but he probably won't be able to hold as high as I can, and so he won't be able to jump on as stiff a pole.

"If everything is going well, if you're having a good day, you go to a stiffer pole. There's an optimum hand hold, an optimum point on every pole that a given vaulter of a certain weight and a certain speed can manage," Olson says. "If he tries to hold too low, the pole won't bend enough.

"If you get a pole that's too stiff, you won't get to the pit. If the pole is too soft, you leave the ground and you're moving so fast and there's not enough pole to stop you, so you land out of the back of the pit." In either case, you miss the pit, which is unquestionably the worst thing that can possibly go wrong in the pole vault.

The Backward Physiology of Crew

JOHN KIEFER

Even at the starting line, one begins to suspect that crew is much harder than it looks. Their long sweep oars motionless in the water, the eight oarmen sit compressed like a giant spring, knees up under their chins and arms stretched to the limit of their reach. At the gun they sprint off the line with a precision and economy of motion that makes their rowing look almost effortless. In just ten strokes, they reach a top speed of about thirteen miles per hour. The oarsmen work together by keying on the aftmost rower; each time he slides forward on his movable seat they follow. The oar blades slip into the water simultaneously, the legs straighten, and the boat is levered forward. The stroke continues as the upper body leans back and the arms are pulled into the chest. If the oar blade were not lifted out of the water at this instant, the boat would carry the oar handle into the rower's gut with a force that could easily hurl him from the boat.

During the first 250 meters of the 2,000-meter race, a topflight college team may row, or more accurately sprint, at up to forty-five strokes a minute. Only then will the diminutive coxswain—who steers the boat and sets the rowing cadence—bring down the pace to thirty-five strokes for the next 1,500 meters. Finally, he urges his teammates back up to forty strokes per minute for the 250-meter rush to the finish line. By now, the effects of the all-out opening sprint are being felt. The oarsmen's oxygen-starved muscles are burning with pain. Only after they cross the finish line, just six minutes after their start, do the spectators see the extent of their efforts. The rowers slump forward over their oars, faces contorted, chests laboring for air.

For physiologists, the opening sprint in crew adds dimension not found in any other sport. "It's the only sport lasting more than two minutes that starts with an all-out sprint," says Fritz Hagerman, director of the Work Physiology Laboratory at Ohio University. "No mile runner worth his salt will sprint the first lap because of the oxygen deficit he incurs by doing so."

From a standing start, the body is incapable of supplying the six liters of oxygen required per minute to fuel such a sprint; that is twenty times the quantity supplied at rest. It takes one to two minutes for breathing and heartbeat rates to increase, blood vessels feeding the working muscles to dilate, and the additional oxygen to diffuse through the capillary walls and into the muscles. While they're waiting for the body to make these adjustments, athletes get their energy anaerobically—without oxygen—during the first few minutes of a race. Anaerobic metabolism is not as efficient as aerobic (oxygen-derived) metabolism, and its chemical by-products, which include lactic acid, cause muscle fatigue and pain. In some events, such as the 100-yard dash, athletes rely entirely on anaerobic metabolism because the race ends before muscle fatigue affects performance. To avoid fatigue in events lasting longer than a few minutes, swimmers and runners, for example, save their sprints for the end, gradually building up their body's ability to deliver oxygen to the muscles.

Rowing's backward approach to metabolic efficiency is the result of two characteristics unique to the sport. Rowing shells have no keel, making them unstable. The wake created by a lead boat can easily rock a trailing boat and disrupt the rowing rhythm of its oarsmen enough that they become momentarily out of sync with each other and lose precious seconds. It is also psychologically important. "Since rowers face where they came from rather than where they are going, getting behind at the start puts a team out of touch with its opponents," says John Ferriss, crew coach at Cornell University.

The need for an initial sprint in a race covering more than a mile is one of the main reasons crew is so physically demanding. But a good rower must have other characteristics besides endurance and well-developed thigh, lower back, and shoulder muscles.

In trying to pick the eight best rowers for men and women crews, coaches have traditionally resisted highly technological means of testing strength and endurance such as computer-generated plots of body movement and monitoring blood chemistry. The one exception is the ergometer, a $500 machine in which the rower can duplicate the rowing stroke without going anywhere. By measuring the number of times the rower is able to turn a flywheel, the ergometer measures a rower's work output under simulated race conditions. Rowers with high ergometer scores (three thousand or more for six minutes) generally have better race times than those with lower scores.

Still, devices like the ergometer cannot measure a rower's ability to function as a member of a team. For the two aftmost rowers, rhythm is as important as strength, since these two rowers provide each side of the boat with the rowing pace. Similarly, the two rowers in the bow need a good sense of balance, because the bow, being both exposed and narrow, is the easiest part of the boat to rock. Only in the center four seats is strength the prime asset.

With the physical training that crew demands, even the most devoted rowers always welcome a little help from their designer friends. In the early 1800s the shells weighed more than 600 pounds, were wide and bulky, and had stationary seats like a rowboat. The most significant advance in boat design was the movable seat, developed in 1870. This seat, which rolls on four small wheels along a two-and-a-half-foot track, greatly increases the length of a rower's stroke and allows him to make much better use of his legs.

Today boat designers are building faster and more responsive boats by making them from new lightweight materials. Boat weight is important because the primary force the rowers are working against is frictional drag on the outside of the shell as it passes through the water. The heavier the shell, the deeper it sinks into the water and the greater the area that experiences drag.

Besides the simple cedar boat with protective fiber glass coating, teams now can choose between boron fiber boats, composite boats made of specialized plastic laminates, and aluminum honeycomb boats. For all this apparent diversity, one new material—carbon fiber—has managed to make its way into practically every one of the new generation of crew shells. Carbon fibers can reduce the weight of a crew shell 40 pounds or more over traditional wood boats, which average a mere 270 pounds as is, without any loss in shell strength or rigidity. Structural rigidity ensures that the force exerted by each rower is efficiently spent. "A boat that is not rigid," says Steve Gladstone, Brown University crew coach, "tends to waste energy by moving or flexing in directions other than forward."

To avoid such flexing while still reducing weight and subsequent drag, boat designers have applied sheets of carbon fibers on top of fiber glass and plastic laminates "like long strips of Scotch tape in the bottom of the boat and from side to side like so many ribs," says Gladstone. Other boats have actually been built as a shell within a shell, both shells being composed of aluminum honeycomb sandwiched between layers of carbon fiber.

Carbon fibers are also being used in the twelve-and-a-half-foot oar that each rower uses. In a new composite oar made of fiber glass and carbon fiber, the only wood left is in the handle, and that's only because rowers like to carve it up to fit their grip. These oars have equal to or greater strength than the traditional spruce ones but weigh as much as three pounds less.

As these innovations help oarsmen go faster, they also force them to become more skilled. "A crew shell is like a water spider balancing on long, spindly legs," says Gladstone. "By lifting its body up out of the water and lightening its legs, you make it more likely to fall over." To compensate for this, oarsmen must more than ever row with a combination of power and finesse—making crew, says Ferriss, as much "a sport of sense and feel" as it is one of strength and endurance.

Straight Shooting

TIM PELTON

By the end of the second day of the 1980 U.S. International Shooting Championships, Lanny Bassham held second place, hitting more than 150 bull's-eyes out of the required 240 shots from 165 feet away. But by the third and final day of the competition, he had slipped to fourth place. Physiologist Daniel Landers knew why. During the first two days, nearly every time Bassham fired, he pulled the trigger right before his heartbeat. On the last day, however, his shots did not match his heartbeat but came haphazardly before, during, and after the muscle's contraction. And the bullets were off target.

It doesn't take much to be off target in this sport. The object is to make a .22-caliber bullet (.22 inches in diameter) punch through a bull's-eye the size of a dime—and a heartbeat's reverberation through the chest can send the shot astray. Nevertheless, top marksmen in a prone position can hit bull's-eyes an average of fifty times out of sixty; on two occasions a shooter hit sixty out of sixty. Such marksmanship requires more than a keen eye and a steady hand.

"If you take five shooters with the same technique," says sport psychologist Frederick Daniels, "the winner will be the one with the best mental training." To find out just what separates the world's best shooters from all the others, Landers, Daniels, and their co-workers at Arizona University have hooked up hundreds

158

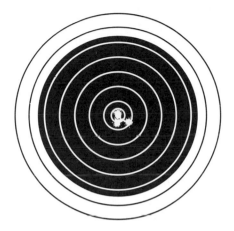

This target bears the work of marksman Mike Anti. Hits on the smallest ring do not score extra points—they are used to break ties. Hitting anywhere within the next largest ring scores a bull's-eye—ten points. To get an idea of how the target appears from a 165-foot shooting range, look at this photograph from about 65 feet away. (© Sharyn Anti)

of shooters to machines that monitor brain wave patterns, heartbeat, the skin's electrical response, muscle tension, and respiration during shooting sessions. It turns out that there are optimum ranges of these functions for each shooter. For example, most marksmen generally have an increase of eight to twenty beats per minute over their normal resting heart rates. Above or below that pace the shots begin to miss.

Landers also found that shooters consistently squeeze the trigger between heartbeats. A perfect shot was rarely recorded if the heartbeat and the trigger pull coincided. Landers believes that marksmen use a heartbeat as an unconscious cue to fire. The lapse between beats is long enough to allow the shooter to squeeze the trigger and allow the bullet to exit the barrel before another beat occurs.

The researchers discovered that just before a shot, a marksman's brain is in its most relaxed state, indicated by a high proportion of alpha waves. Beta waves, which occur during periods of high mental activity, are significantly reduced. Some shooters, in fact, have used drugs—now outlawed in all competition—that slow the heartbeat and block these beta waves.

To discover whether shooters relied on intuition and mental images from the right side of the brain or on logical and factual information from the left side, the team compared electrical signals from the two hemispheres. They found that in most excellent marksmen the right hemisphere sustained its activity during a shot while the left side became increasingly relaxed, indicating that the shooters rely on getting a proper "feel" rather than logically thinking about their motions during the shot.

According to the International Marksmanship Guide, as a marksman prepares to shoot he should gradually take deeper breaths, then hold the breath before, during, and after the shot, and then inhale again before exhaling. But the researchers found that the best shooters breathe evenly before the shot and finish by exhaling. "Some shooters gradually take smaller breaths before they shoot," says Daniels, "but nobody takes deep breaths."

Though many rifle shooters have vision better than 20/20, Landers says that the finest pistol competitors, by contrast, are extremely nearsighted. Some even wear shooting glasses that cause them to be nearsighted. Apparently, says Landers, it is more important to keep a handgun's sights in near-perfect focus than to focus on the target.

Research on vision turned up other important evidence. Just as everyone has a dominant hand, most people rely on the input from one eye more than the other. According to Landers, shooters with mismatched hand and eye dominance could improve by switching hands: A right-handed marksman with dominance in the left eye should shoot with his left hand.

The goal of the researchers' work is to train shooters to control their body rhythms by using their minds. In competition, shooters generally get seventy minutes to go through sixty targets. If a shooter finds that he is not concentrating or relaxing enough, the researchers prescribe meditation between shots. "What works for you may not work for me," says Daniels. "It might be the image of walking along a beach or lying in the grass. Some shooters think through the image of themselves performing a perfect shot." Others might try to tense and relax individual muscles.

Some shooters have learned to control their heart rate and brain waves by hooking themselves up to an alarm system. If the heart rate goes out of its optimal range, the alarm goes off. The shooter then concentrates on returning the heartbeat to that range. During a match a competitor uses the same technique, only without

HEARTBEAT

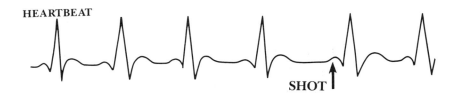

SHOT

This .22-caliber rifle, the most popular among competitive shooters, has an adjustable stock and handgrip. Telescopes are not allowed. Its most important feature is the trigger mechanism. The time between the pull of the trigger and the ignition of powder in the shell—crucial because during that fraction of a second the shooter can move the barrel off the target—is the shortest of any rifle. The chart shows the heartbeat of marksman Lanny Bassham during the first few days of the 1980 International Shooting Championships in Phoenix, Arizona. Bassham's heart rate slowed as he prepared to shoot, which helped him fire right before the "spike" of his heartbeat. (Top– © Frederick S. Daniels and Daniel Landers. Bottom– © Anschutz Rifles)

the alarm. "The electronics allows the shooter to know what's going on inside," says Daniels. "They're impressed to see it, and even more impressed when they find they're able to control it." By using such biofeedback techniques, Landers' team has helped some shooters raise their scores by as much as a critical 5 percent.

Landers says that shooting requires no special body strength or size. Thus women can compete on equal terms with men. "Physiology is not a problem for women," says Daniels. "It might even be easier for them. One type of shot requires the shooter to rest the arm on the hip, and a woman's larger hips acts like a shelf. And they generally have a lower center of gravity, which is another advantage."

In fact, many top shooters are women, says Daniels. Pat Spurgin of Billings, Montana, took 4th in the 1983 U.S. Junior Nationals, and Margaret Murdock of the U.S. Olympic team won a silver medal at the Montreal Olympic games.

Such mixed competition, however, may become a thing of the past. The Olympic Committee has decided that for the 1984 games shooting events will be separate for women and men—a move that has many U.S. shooters and the National Rifle Association up in arms. The women's events will also require fewer shots, which, according to the NRA's David Boyd, is intended to prevent comparisons between the sexes. "The male chauvinism of the Europeans is overwhelming," says Boyd, a former world-record holder. "When a European loses to a woman he takes it personally."

Marathoning with Maps

LINTON ROBINSON

You're standing in the middle of a wooded grove far from any road or house. There's a compass in your left hand and a detailed map in your right. A series of checkpoints is marked on the map, and you estimate that the next one is about a half mile away. But a group of concentric contour lines drawn in brown tells you there is a steep hill between you and your goal. The route over the hill would be shorter, but would it be faster? Would it be better to save time by climbing now, while the legs are still fresh? Or should you go around the hill, following the stream marked in blue on the map? And is it wiser to cross the stream now or later?

This is orienteering, a mixture of marathon, hike, and scavenger hunt, a cross-country race in which participants must locate a series of markers set in unfamiliar terrain by means of map and compass. The course, which may range from an acre of city park to twenty square miles of wilderness, is dotted with anywhere from four to fifteen "controls," red-and-white flags whose general locations are marked on the map by small circles. At each control there is a paper punch that produces a distinctive pattern on a card the racer carries. In most events the order in which the card must be punched is fixed; the route taken to reach each control, however, is up to the participant.

Beginning at staggered times to avoid crowding, runners are clocked from the moment they receive the map until they cross the finish line. Novice routes, termed "white" under a standard international system, can be as simple as a one-mile walk in the park with occasional stops to get one's bearings. A more demanding "blue" route, on the other hand, might entail climbing steep ridges, circling around swamps, and fording rivers. Eighty-minute times are not unusual for winners of an eight-mile blue route, though stragglers might end up taking all day; errors in map reading can be costly.

While the growing ranks of American orienteers are filled with long distance runners and nature lovers attracted by the chance to develop pathfinding skills, the sport also seems to attract problem solvers, people with an analytical bent. Many orienteers are engineers, computer programmers, and lawyers. "Orienteering is 50 percent muscle and 50 percent mind," says Robert Defer, director of the United States Orienteering Federation, which coordinates national meets. "It doesn't matter how fast you go if you're moving in the wrong direction."

Most orienteers use an ordinary plastic compass to align the map with the surrounding terrain. The orienteer then uses his wits to compare the map with what he sees. One technique is to look for "handrails," features such as a stream or the edge of a forest that lie in the same direction as the next control point. While sprinting along handrails, orienteers also look for objects such as houses and boulders—which appear on the map—to gauge how far they have traveled. Some even count paces to estimate the distance covered.

The most exacting exercise connected with the sport is not done in running shoes. Orienteering maps, made by specialty services in Norway, Scotland, and Germany, carry the art of cartography nearly to the level of obsession. While U.S. Geological Survey maps might be adequate for hikers and soldiers, orienteers need more accurate and detailed information. To produce them, two aerial photographs are taken of a course from slightly different angles. This simulates the effect of seeing the area with two large and widely separated eyes. The two photographs are formed into a three-dimensional image by a photogrametric plotter, a machine that operates somewhat like a stereo viewer. The machine also superimposes a small white dot of light on the image. The mapmaker can make the dot appear to be at any elevation above or below the ground. After "lowering" it to a location whose elevation is known, the mapmaker carefully moves the dot along the terrain at that level. The same controls that move the dot are connected to a mechanical arm that traces the motion onto a map.

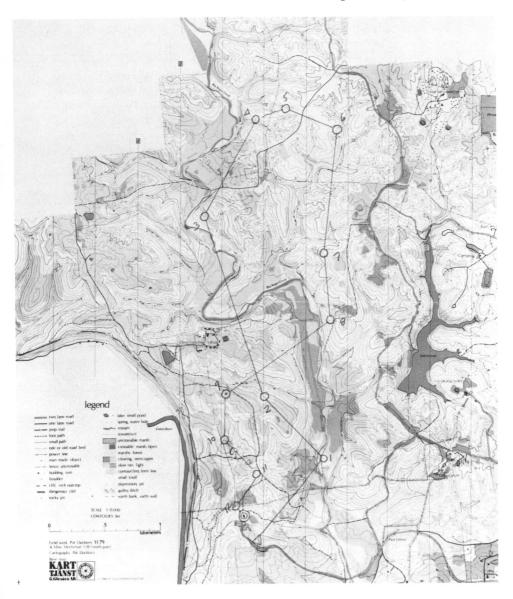

This orienteering map of *Cuivre River State Park* in Missouri shows the landscape in painstaking detail, down to uncrossable fences, power lines, and springs. The distance between each contour line is about eighteen feet, and a group of concentric lines ending in a small circle indicate a high hill. Though the numbered markers must be found in order, the route used to get there is up to the runner. (© St. Louis Orienteering Club)

By setting the dot at different elevations, a mapmaker can draw a series of contour lines over the course. Internationally recognized orienteering symbols are then inserted to indicate trails, buildings, rivers, and sometimes even individual rocks and specific kinds of trees. This is still not exact enough. After a rough map is made, ground crews tour the area, checking the map and adding more details. Orienteers boast that their courses are the most carefully mapped areas on the planet.

If this sounds like cartographic overkill, remember that the orienteer, unlike the hiker or soldier, is not looking for a particular meadow, river, or mountain, but for a control that may be hidden behind a specific boulder or stump. And he is trying to do so while moving at racing speed.

But it's not necessary to run at a breakneck pace to enjoy orienteering. Many grade schools have orienteering programs to help teach biology and mathematics. "You show students how to judge the distance they've traveled by measuring one of their paces, converting it to meters, and converting that to millimeters on a map, and before they know it you've taught them some math," says Defer.

Orienteering has a large following overseas. It's already one of the most popular sports in Scandinavia, Central Europe, and Japan. In Sweden, where orienteering ranks with soccer as the national pastime, a sixth of the population is navigating the countryside on any weekend. International championships are hotly contested between Scandinavians, Swiss, and Czechs.

Orienteering is catching on fast in the United States, too. Many major cities have orienteering clubs, and the National Championship is expected to draw more than 500 entries. One of the reasons for the sport's popularity is that it is open to many variations. Different courses are designed for bicycles, horses, canoes, sailboats, and wheelchairs. And for the hard-core orienteer, there are even orienteering meets on skis that require a flashlight as well as a compass—they're held at night.

The Finer Points of Darts

JOHN TIERNEY

One recent evening Roger Nickson, the governor of the Morning Star pub in London's Peckham district, sat drinking ale and trying to explain why he is England's most successful darts manager. He was in a second-floor room with flaking paint and holes in the wall at the Walmer Castle, another pub in Peckham, and his team was doing its usual job of demolishing the opposition. Nickson refused to take credit.

"Darts is not a game where coaching goes a long way," he explained as one of his players, a truck driver named Dave Willingham, prepared to throw. Willingham was not your model athlete. He was at least fifty pounds overweight and had already consumed more than a couple of ales. He threw the dart with a peculiar stutter in his swing—a brief hesitation midway, then a jerk forward. His first dart landed on 20, barely outside the tiny triple-bonus area that he was aiming for.

"He does it all wrong," whispered Nickson. "Did you see how he lifted his back foot? That shouldn't move."

The second dart landed inside the triple 20. So did the third, prompting a chorus of "Lovely darts" from the crowd.

"All wrong," repeated Nickson. "And you see what he gets— 140. It makes you wonder if coaching makes any difference. Sometimes I think my main job is to make sure my players don't drink too much."

Darts is that kind of game. It is the simplest of sports—one brief motion repeated over and over—and at the same time one of the least understood. Experts argue about how to stand, how to hold the dart, how to throw it, what to aim for, what kind of dart to use. A few decades ago the world's largest dart company, Unicorn Products, Ltd., in London went to physicists in a British government laboratory and asked how to design the perfect dart. The scientists obliged with a few tosses and calculations, but no conclusion.

"It seemed like an elementary exercise in ballistics, but the physicists told us that there was no such thing as a perfect dart," recalls Stanley Lowy, who runs Unicorn along with his brother. "They said there was too much variability in the throws and the trajecto-

On this standard "clock board," the top dart scores one point, the bottom dart in the double-ring scores four, and the dart in the inner triple-ring scores twenty-one. Most players aim at the highest score, the triple-twenty. But a miss there can be costly—a one or five—so some aim their throws at sixteen or nineteen. (© Rogers and Gahan/PRISM)

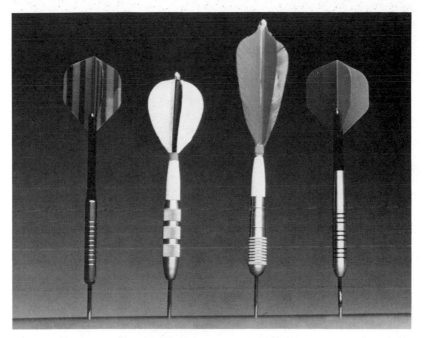

These four darts all weigh between twenty and twenty-two grams. The center two have the traditional brass barrels and turkey feather flights. The other two darts, with polyester flights, have slimmer barrels made of denser metals: a mixture of tungsten, iron, and nickel at far left and a silver-nickel alloy at far right. Each dart's shaft—the section between the barrel and the flight—is plastic. (© Rogers and Gahan/PRISM)

ries people use. We've found the same thing here in our reseach department. We've built throwing machines to test darts, but we don't find them terribly useful because while we can reproduce an average throw, we've never found a player who has an average throw."

All the uncertainty, of course, hasn't stopped anyone from trying to build a better dart. The quest has been going on for at least a century and probably much longer. Some historians argue that spear-throwing Stone Age warriors were the first dart players (as well as the first targets), although the more conventional view is that medieval archers started the sport by throwing their arrows at the end of a log. The log, with its concentric rings and cracks emanating from the center, is presumed to be the ancestor of the standard "clock board," which was designed at the end of the nineteenth century. In those days the players used a single piece of wood attached to the point, with feathers or a paper flight at the other end. By the 1930s the main body of the dart had been subdivided into a brass barrel up front—the extra weight improved accuracy—and a wooden shaft in the rear to hold the flight.

This arrangement, however, didn't satisfy a Hungarian engineer who was visiting England to sell boiler linings. In 1936 he designed the first all-metal, precision-made dart, consisting of a brass barrel and an aluminum shaft. The engineer, Frank Lowy, called his dart the Silver Comet and started the Unicorn company. His sons now sell five hundred kinds of barrels, twenty-one kinds of shafts that attach to the rear of the barrel, and six styles of feather and plastic flights that fit the shafts. They can be variously combined to produce thousands of darts of different weights, shapes, and sizes within the permissible limits—no more than twelve inches long or fifty grams in weight. The traditional brass barrels are still the best sellers, but in the past decade most of the top players have switched to tungsten alloys. Tungsten is twice as dense as brass, so the tungsten barrels are slimmer, which makes it easier to land the darts in the same area—"three in a bed," as it's known.

"I don't know that we're going to find any better metals than tungsten," says Stanley Lowy. "You can look at the periodic table and find similar metals that are even denser, but they get so expensive that it just wouldn't be worthwhile. I suppose you could use uranium, although I'm not sure the public would feel safe with it."

Tungsten's rise in popularity has been accompanied by the decline of the feather flight. It's made from the feathers on the leading edge of turkeys' wings. Because each quill is so thin and flexible, it's easy for a dart to pass through the feathers of another dart already on the board. After the round it's simple to brush the feathers back in place. "There's no synthetic material that we could make with such narrow, resilient sections," Lowy says. "Aerodynamically, it's probably still the best flight made." But turkey feathers are expensive and they're easily damaged, particularly when hit by the dense tungsten darts. Players also complain that they've become weaker now that turkeys are raised in confinement instead of out in the barnyard. So most darters, including the professionals, use the cheaper and more durable polyester flights. Most also use shafts made of plastic, although some darters prefer cane, and a few have begun using titanium. A titanium shaft is expensive, but it's very thin, durable, and can bend slightly to let another dart pass.

Some inventors have offered even further refinements. One designed a dart with a spring behind the tip to help absorb the shock when a dart hit a metal divider on the board—the theory being that the dart would then slide into the board instead of bouncing

When the dart's nose points downward, the upper side of its feathered tail meets air resistance. This pushes down the tail and points the nose back toward the horizontal. (© Rogers and Gahan/PRISM)

off on to the floor. Another built a dart with a barrel, shaft, and flight that were supposed to fall off when the dart hit the board; this would leave just the slender point imbedded in the target, making it easier to get three in a bed. Neither dart, however, is legal in tournaments.

Once you've chosen a dart, you face two major decisions: how to throw it and how much to drink. Scientists don't offer much help on the first one. Stanley Mohler, a Dayton, Ohio, specialist in aerospace medicine, took time off from his studies of pilots to perform what is probably the only physiological analysis of dart throwing. His conclusion is that you should do what's comfortable. For most right-handers, he says, this means standing with the right foot forward. "By leaning forward slightly and putting a twist of about forty-five degrees in the spine," notes Mohler, "additional stability to the torso is provided." To make the body a stable "launching platform," Mohler says, the lower arm should be the only part of the body to move during a toss.

There's no scientific literature on drunken dart players, but there are some relevant results from intoxicated rats. A recent study has shown that an animal can perform some tasks reasonably well while intoxicated—but only if the animal has practiced that task while intoxicated. This suggests that practicing darts while sober at home may not be too useful for a game at the bar. Conversely, some studies have shown that if an animal learns a task while drunk, it later performs that task better when it's drunk than when it's sober. This theory of "state-dependent learning" suggests that a regular player at a tavern should think twice about going into a tournament sober.

As to whether alcohol in itself helps a player's game, that's debatable. Researchers have found that a tiny amount—less than a quarter of a shot—actually improves people's reflexes, presumably because it eases distracting tensions. Greater amounts start slowing reflexes, although it's not clear how much this hurts a dart player. In some tense situations, Mohler says, any ill effects of a beer can be more than offset by the relaxation it brings.

"For a complicated task, like flying a plane, alcohol's effects show up immediately. But the basic skill of throwing a dart is rather elementary. There's no opponent attacking, no floor moving underneath, no object coming at you. You're dealing with a lower-order reflex, which probably isn't seriously affected by the alcohol in one or two beers." On this point Roger Nickson agrees.

"I do have one player who only drinks lime juice during matches," Nickson explained at the Walmer Castle. "But that's only because he has a long drive home. As a general rule, a dart player drinks. I don't mean that he should get drunk—that doesn't help anyone's game. But I'm a believer in having a pint."

Index

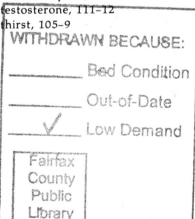